THE WEATHER IS FRONT PAGE NEWS

This work is dedicated to the following individuals:
Gordon Salvetti, who gave me courage;
Louis B. Pomeroy, who gave me strength;
and Sandra, who gave me love.

THE WEATHER IS FRONT PAGE NEWS

Ti Sanders

ICARUS PRESS
South Bend, Indiana

Manufactured in the United States of America.

Icarus Press, Inc.
Post Office Box 1225
South Bend, Indiana 46624

1 2 3 4 5 86 85 84 83

Library of Congress Cataloging in Publication Data

Sanders, Ti, 1947-
The weather is front page news.

Includes indexes.
1. United States—Climate—History. 2. Natural disasters—United States—History. I. Title.
QC983.S26 1983 973 83-18511
ISBN 0-89651-902-3
ISBN 0-89651-903-1 (pbk.)

CONTENTS

FOREWORD

THE BROAD FLOW OF INFORMATION FROM PEN AND PRESS GOT UNDERWAY IN AMERICA shortly after the turn of the nineteenth century, and by 1850 reports from various parts of the United States were available in the form of daily newspapers, including the New York *Times*. In writing this book, which constitutes an effort to highlight some of the more spectacular weather events as they occurred across the nation from the mid-1850s to the present, I was able to glean much information from that publication, as the *Times* was one of the first newspapers to include correspondence from all sections of the country. In some instances, to be sure, I availed myself of other sources for good local accounts.

When the Weather Bureau was established in 1871, a systematic effort was put into effect by the Government to standardize weather observations and forecasts and to compile data on the many vagaries of the weather. The tools for such endeavors were limited when compared to the vast arsenal of computers, weather satellites, and other electronic equipment available today. Only a generation ago, scores of meteorologists were employed to provide a daily weather map. Today that is the work of the computer, based upon the system of numerical modeling, and four complete forecasts are prepared daily from thousands of bits of information collected from around the world. In a typical large-city office, only two meteorologists are on duty at any given time to interpret the data.

It is interesting to view outstanding weather phenomena in the context of the period in which they occurred. Witness the great Blizzard of 1888, which began as a heavy downpour of rain after a spell of fine spring weather. Forecasters were as astonished at the results of that storm as they are today by the precarious heat wave and drought of 1980. While the instruments exist to promote the fine tuning of a modern-day forecast, accuracy levels drop sharply to a point of guesswork beyond a period of two or three days. In other words, despite our advanced technology and the fact that short-term forecasts are far more reliable than they were a century ago, it is Nature which has the last word, and any of the disasters described in the pages of this book can happen again and are *liable* to happen again. Preparedness is the key to saving lives.

While efforts have been made in the field of weather modification, as in the hurricane seedings of the early 1960s, results have not been forthcoming. The dynamics of the weather may be better understood today, but the comprehension of a hurricane's mighty force—about the equivalent of 10,000 megaton atomic bombs of energy dispensed every hour—make it a formidable foe. Rainmaking in drought-prone areas has met with some success, but results are often scattered, and there is no way of proving that the rain might not have fallen had nature taken its course. Weather modification, in short, is a science in the embryonic stages, and it could well be that even a thorough understanding of the physics of the atmosphere will prove to be of little benefit.

Thus, we are faced with a need to study past conditions, to understand the limits of each season, and, in our astonishment, to be prepared for the mightiest of onslaughts that Nature can provide. While an effort has been made to choose representative examples of each of the different weather phenomena and to discuss the geographical extent of their influence, this is by no means a complete compendium of our meteorological history. Rather, these events are highlights from a variety of observing stations with events selected randomly from approximately 130 years of American meteorological annals.

One of the most intriguing aspects of this research was not the blow-by-blow description of an infamous flood or blizzard but rather a look at the human perspective. Individual accounts of survival are sprinkled generously throughout the text in an effort to put the reader at the center of the maelstrom, so to speak.

Weather, to a large degree, has shaped the culture, folklore, and to an extent the destiny of this country. For example, on Christmas Day, 1776, General Washington's troops were surrounded by the British at Valley Forge, and an attack was imminent. Washington was an astute observer of the weather in all of its manifestations, and he correctly noted the passage of a cold front, knowing that the mudflats upon which his men were stranded would freeze solidly overnight. Setting campfires to betray their presence, Washington and his troops safely escaped Cornwallis and the British under cover of night.

Tornadoes, floods, hurricanes, and snowstorms. Each has its season and has left an indelible mark on the American landscape. We all are affected by the weather, each and every day, in striking and in subtle ways, and it's only a matter of time before weather is on the front pages once again.

I. COLD WAVES

A cold wave in mid-winter in the United States is often the source of some of the most dynamic atmospheric antics and discomfort of any phenomenon that the seasons offer. Perhaps it's due to the increased fuel costs they bring to the northern states, or the agonizing winds that accompany the frigid Canadian blasts or the epidemic influenzas that are associated with a long cold spell. Whatever the reason, they are frequent intruders from the hinterlands of the North and, as Bob Dylan observed, "you don't need a weatherman to know which way the wind blows," especially if it's a cold wind in January. No portion of this country is completely immune to the sudden drop in temperature from an advective winter blast, including the Sunshine State, where temperatures in Miami have descended to twenty-eight degrees and snow was flying through the air in January of 1982.

The actual "cold front" is the concluding wave of a migratory storm system, and marks the boundary between a retreating mass of milder air (which is often accompanied, in winter, by winds from a southerly component, rain or snow) and the invasion of a fresh blast of cold air. The core of the cold air frequently lies thousands of miles to the west of the frontal passage, although the front's passage usually brings the most dramatic change in temperature and wind direction. During the summer months, when storm systems track well to our north, through southern Canada, the trailing cold front can spell relief in a cooling thundershower and the introduction of a pleasantly cool and dry air mass for a day or two before it is modified by the high summer sun. Refreshing as a change of air masses is in mid-July, the January winds from the north can spell misery, even catastrophe following the December solstice.

Depending upon the origin and trajectory of the winter air mass, we measure its effects in terms of cold, colder, coldest. An air mass traveling across the country via the Pacific may hardly be noticeable, as the long journey over land has permitted the air mass to moderate to seasonable levels. In the northern-tier states of Minnesota and North Dakota, a fresh blast of polar air arrives relatively unmodified from its source, as the snow-covered terrain reflects much of the incoming radiation back into space without retaining much warmth. Temperatures under these con-

ditions can tumble to thirty-five to fifty degrees below zero, with much of the descent taking place in a matter of hours. Similarly, the New England states are prone to such blasts, particularly if the air mass passes to the east of the Great Lakes from Hudson Bay, a distance of only 400-500 miles. Cold air masses passing over the Great Lakes themselves advect the moisture from the relatively mild waters and produce copious snowfalls on the lee sides of the lake, as was the case in Buffalo during the winter of 1976-77. Their passage through the heart of the country can mean moderation by the time they slide off the East Coast.

The position of the jet stream in winter often holds the key as to which sections of the country suffer most. During the extreme cold of 1976-77 in the East, western states witnessed some of their mildest winter temperatures ever known. The jet stream, which sometimes resembles a coiling snake, thrust sharply to the north off the Pacific coast, bringing milder air to Alaska before its abrupt passage southeast, becoming colder across the frozen tundra of the Yukon and Canada. Many portions of the United States set all-time minimums and experienced their coldest January since 1780. Cincinnati experienced temperatures for the month of eighteen degrees below normal while many other cities east of the Mississippi River were between five and fifteen degrees below normal for the period.

When the cold front reaches the warm moist air of the Gulf of Mexico, the interaction between air masses produces some of the mightiest of winter storms, which track northeastward through the Ohio Valley, or make a run up the East Coast. This in turn pulls down another blast of arctic air, renewing the seemingly endless procession of Canadian air masses. Often, it is the second or third night following the passage of a cold wave when the coldest temperatures are experienced. If there is already a blanket of snow on the ground, with skies clear and winds calm at the center of the air mass, conditions for thermal radiation are ideal, and it is under these circumstances that many all-time-low figures have been established.

Historically, there have been only a handful of genuine zero days in the East, where temperatures have remained below zero for an entire calendar day. Two such days occurred in the middle of the nineteenth century in what H.D. Thoreau was to describe in his diaries as the "long snowy winter." January 18-19, 1857, sent temperatures in Fort Ripley, Minnesota, tumbling to historic lows of -50°F or lower, while Boston's mercury stood at -5.5°F at 2:30 P.M., the lowest early afternoon reading ever known in the Yankee City. Three days later, a second and even colder air mass descended upon the United States, erasing some of the shortest-lived standards in the record books. Perhaps one of the coldest air masses ever to descend on this country occurred in the Great Arctic Outbreak of 1899. All-time-low readings were set from Montana to Texas, and east to the Atlantic coast. Absolute state minimums as of that time were set in twelve southern

Following an intense winter storm, Arctic air is frequently drawn southward, creating a wintry scene like this one. (*Ron DerMarderosian*)

states as the thermometer at Tallahassee, Florida, sank to -2°F, the lowest ever recorded in the Sunshine State. A giant sprawling blizzard of the same dimension as the Blizzard of 1888 was spawned as a result of this air mass and is discussed in the chapter on snowstorms. The coldest winter month in the Northeast was in February of 1934, although January of 1977 subsequently toppled those records. The two winters that followed each had their claim on frigid records, and they stood until January of 1982 when the ten-day period from January 9-18 toppled many more all-time-minimum records throughout the Midwest (Chicago: -26°F) and East, where yet another zero day was compiled.

The following examples illustrate some of the many human miseries that can accompany a cold wave, especially in times of war or fuel shortages. Oftentimes it is the cold air alone that causes most of the damage and takes the greatest human toll of lives; yet, just as many of the problems of a heat wave in midsummer can be compounded by the scourge of drought, the treachery of snow often accompanies winter's bitterest conditions.

1. SCENES FROM AN OLD-FASHIONED WINTER

THE METEOROLOGICAL HIGHLIGHTS OF THE MIDNINETEENTH CENTURY OCCURRED IN SOME spectacularly wintry weather which Henry David Thoreau was to dub the Long Snowy Winter; he so entitled that year's volume from his famous *Journal*. The winter of 1855-56 rated near the top among severe winters in the nineteenth century, and in terms of cold, exceeded all other winters. The first substantial thaw held off until the second of April, melting the snowcover that had provided excellent sleighing since the turn of the year. It was not until January of 1857, however, that easterners were to see the worst that Nature in a wintry despond had to offer. That month is singularly rated as the coldest of the nineteenth century, from the Midwest to the Atlantic Seaboard.

Not only was the general average lower than in any other winter month, but extreme minimums were commonplace over a wide swath of the country from Cincinnati to Philadelphia, from Atlanta to Bangor. Temperatures had already touched zero or dipped below in most sections of the East, and a strong coastal storm had left a blanket of white nearly 10 inches deep in most sections, north of the Mason-Dixon line which persisted until the end of the month. These conditions, however, were only a prelude to the fierce snowstorm and two spectacular cold waves that made this month celebrated.

The first was to swarm in upon the Northeast on the eighteenth of January and dropped readings in Upstate New York to minus forty on an average. By Sunday morning, the temperature in New York was hovering near zero, and it was apparent that the wolves of winter came down upon Gotham in earnest. The East River and the Hudson were passable on foot; as the *Times* observed, "Hundreds of both sexes availed themselves of this novel mode of crossing, thereby defrauding the Ferry Company of a considerable number of pennies, which they will receive, if at all, by an action of law." The thickness of the ice varied from one to three feet, and a flow was brought up against a shelf of thicker ice in the East River on an incoming tide. Needless to say, it presented a great temptation for the adventuresome as hundreds rushed to the wharves to enjoy the pleasure and dangers of crossing the river on an ice bridge. A contemporary account described the scene: "Not gentlemen alone, but Ladies—with and without crinolines, youths with and without protectors, representatives of all classes but the sensible." Meanwhile, the outgoing tide made itself known, and hundreds made a life-and-death dash for the safety of the shore while onlookers shared in the excitement. Fortunately, there were no drowning victims, but New Yorkers proved once again what a fun-loving people they are.

Meanwhile, a storm was taking shape in the Louisiana Gulf, and already an unusual snow had broken out in the central sections of South Carolina in what would prove to be the South's most devastating winter storm of all time. Reports up and down the east coast indicated that the sun did not show itself throughout the day, and by midday, snow visited as far north as Philadelphia and New York. The storm in the South is particularly interesting. The storm, which generated rain and thunderstorms in Florida, changed over to snow in the uplands of South Carolina, where the polar air from the north had pressed down sufficiently. The town of Aiken picked up two inches, but a correspondent from Athens, Georgia, described the fall as 8 inches. Over most of North Carolina, except along the immediate shoreline, precipitation fell mostly as snow, and up to 16 inches on a level fell in Chapel Hill, at the University of North Carolina. In Virginia, an unprecedented two feet fell as temperatures slumped to only three above in Alexandria, across the Potomac from the Capitol. What made this storm so unusual was that the South had never experienced such snows accompanied by such a biting wind and temperatures so amazingly low. By the morning of the nineteenth, temperatures had sunk to all-time lows in Jacksonville, and pool ice was reported to be two inches thick, as the bitter cold had effectively invaded the Southland.

Up north, where people theorized that "it was too cold to snow," the clouds thickened and lowered as the snow spread northward and a biting, gale-force northeast wind set in. Newspaper writers aptly described the storm as a winter-hurricane, as it did have a well-defined eye where the sun shone and the snow tapered off for an hour or two before resuming. In addition, hurricane-like damage accompanied the storm as chimneys were blown down, outdoor "widow walks" torn from the frame of their buildings, and weather vanes torn from the steeples.

In Massachusetts, "So furiously did the gale turn the windmill on the property of Stephen Smith that the moving parts caught fire from friction and the entire structure was consumed by flames at the height of the storm."

A dispatch from Philadelphia reported on the evening of the eighteenth that "a terrible snowstorm commenced at noon—snow on the level is six inches and two feet in drifts." That was when they were signing off for the night. Throughout the night, billions of snowflakes were to descend on the entire Northeast, with temperatures at or near zero. Laconically, the New York *Times* was to observe, "railroad traffic will be impeded for some time to come." They commented in an acid tone how the Long Island Railroad would suffer little impact from the storm as they had been put out of commission by an earlier December snow.

The storm carried on at random, and declared itself unaccountable. Just as it pleased, it played with equestrian and pedestrian alike, blew around corners on a journey of discovery and "dived down into poor widows' cellars, like the soft ghostly knock of their sons lost at sea." The oldest inhabitant was consulted, who, as the account reckons, though not as wise as Mr. E. Meriam, was yet qualified to speak as one having authority:

> There is probably no record for 50 years of so severe a snowstorm as opened upon us yesterday afternoon accompanied by intense and unexpected cold. The least intense cold of the day was plus 4 and the wind from the northeast began to blow violently.

> For ourselves, we cannot say that such a storm has prevailed before in our recollection. But combined with an equal depression in temperature, we can scarcely remember such a severe and protracted snowstorm as we are now experiencing.

As in other winter storms to follow, the sound of telegraph wires snapping, crackling, and flashing prevailed, like the noise of wood when the fire is burning briskly. The people who managed to sleep that night through the banging of loose blinds, the rattling of windows, and the crackling of skylights opened their eyes upon an old-fashioned northeaster in all its strength and glory, and those who ventured out undoubtedly put their feet into it.

By Monday night, at least in the mid-Atlantic region, the clouds had dispersed, the stars beamed brightly forth, and the vault of the heavens was clear as crystal. The frisky wind in the streets gave every indication of a crisp dry time ahead.

Washington, D.C. wound up with 18-24 inches, with drifts of four feet, Baltimore had two feet on a level with drifts of six to ten feet, while Upstate New York, New England, and Canada checked in with somewhat less snow but temperatures ranging from minus twenty-seven at White River Junction, Vermont, to minus forty in a spirit thermometer nailed to a barn wall in Watertown, New York. Stations in Ohio reported no snow at all but temperatures as low as minus eighteen, while Chicago shuddered under a relatively balmy minus sixteen. Never before had so vast an area with so many people suffered through such extreme conditions as this.

Throughout the East, it was indeed time to "lay off, to banish all care for business, because business of all kinds was knee deep in snow and nobody

"Shovelling Out," by Winslow Homer. (*New York Public Library*)

could dig it out; to feed on fancies—to make touching poetry, if you had been a poet, or to indite feeling prose, if you wrote prose and had feeling."

Next day when the sun rose, the day was "delicious, clear, sharp and crispy. The sun came up, rather late but cheerful, with his face washed and a smile upon it. He threw back the clouds that impeded his gaze and smiled beneficently on the earth with a round large jolly rubicund mirth-exciting face." Next day, the head of the family lingered before breakfast long enough to digest the better part of the paper while the steak was frying, while the eggs were boiling, while the buck wheats were browning, while the snow was settling. The man of the house would call to the stables to prepare the cutter and the teams while the wives and children put on their furs, and soon their eyes were bright and noses red, driving into the teeth of the wind.

There were, of course, snowballs to be encountered, and some bonnets were smashed, some eyes banged up, and some tempers were spoiled. The scene is reminiscent of Dylan Thomas's *Child's Christmas in Wales*; as the New York *Times* editorialized, "We hope the police will do their duty—for putting an eye out in fun does not preserve the sight, and a pretty lady's face is none the less a ruin for being broken in by a snow ball. Arrest the villainous rowdies who throw snow balls!" The task of digging out usually fell upon the young clerks of the shops whose duty compelled them to encounter the absurdly high drifts in front of the stores, "and it nearly broke the hearts of the young gentlemen and completely broke their backs."

Following the passage of this deep cyclonic circulation, a second more severe Arctic outbreak from Hudson Bay was mustering its forces in Quebec; it was to produce the coldest day of the century in most parts of the Northeast, January 23. A brief thaw ensued between the Great Cold Storm and another coastal disturbance, which passed on the twenty-first, and bitterly cold air plunged into the trough behind it, sending temperatures well below zero throughout the region. Boston experienced a genuine zero day on the twenty-third, with temperatures remaining below zero even in daylight. Temperatures of minus fifty or lower were recorded throughout the northern New England states, records which even today are unequaled. The extreme cold can well be measured by the isolated condition of Nantucket Island at the extreme southeast portion of New England, and normally not subject to extremes in temperature. The waters of the sound between it and Cape Cod sealed off solidly on the fifth of January as the first severe cold wave of the month passed, and no schooner could overcome the ice blockade to deliver mail to the island for nearly a month. Temperatures had never before plunged below zero on Nantucket, yet on the morning of the twenty-third had plunged to -6.5°F.

Winter was not over yet. In April, two snowstorms struck the Northeast, particularly in the mountains of Pennsylvania and New England, which left another mantle of white nearly three feet deep on the thirteenth of the month. A second snowstorm on the nineteenth and twentieth affected the same area that had been visited by phenomenal amounts only six days earlier. While eastern sections had rain from this disturbance, "over three feet of heavy wet snow fell on 20-21 April, to raise the seasonal total to 144 inches," or four feet above normal at Cummington in the Berkshires.

That the cold terms of 1855-56 and 1856-57 occurred at all is worth mention, but the back-to-back nature of these two phenomenal winters in the middle of the nineteenth century lent meteorological substance to an otherwise lackluster era.

2.
THE COAL FAMINE OF 1917-18

THE WINTER SEASON OF 1917-18 GOT OFF TO AN IMPRESSIVE AND TIMELY BEGINNING AS A large sprawling snowstorm enveloped the eastern two-thirds of the nation. With a thick blanket of fine powdery snow laid down, the stage was set for a remarkable wintry siege of weather which lasted well into February of the following year. In fact, in the latter days of January and right through the end of the winter season, the nation was to come to a virtual standstill owing to the elements. The snowstorm which struck on the winter solstice was the first factor, of many, to cripple the nation's coal supplies, halting already sharply curtailed deliveries due to America's only winter involvement in World War I.

The northern tiers of the country from the High Plains to New England were the first to experience what has been described as the most frigid and prolonged blast of Arctic air of modern record. Readings on Christmas Day were in the balmy forties and fifties throughout the northeastern corridor, but by the following morning, the mercury had dipped to minus six in Burlington, Vermont, heralding the first of ten consecutive below-zero mornings, with no daytime readings above 32°F until January twelfth. By the twenty-ninth, the cold had pierced all of New England, and the official reading for Boston established an all-time low of -14°F. Other locations in New England were reported to be in the unimaginable minus forties, while New York City checked in with minus six. The newspapers noted that Gotham was "just two jumps ahead of actual need" in terms of coal supply. New York City was especially hard hit by the turn of events to come, for there was inadequate storage space for coal to satisfy a city of that enormous population, and only a five- or six-day supply on hand. By the thirtieth of December, the mercury fell to a dismal -13°F, the coldest day on record in New York until February of 1934.

Temperatures throughout the Midwest and the mid-Atlantic states remained below zero all day on the thirtieth, an unusual phenomenon in itself, but to compound their misery, great gobs of snow fell from the sky, tying up transit and putting a pinch on the scarce fuel supply. The last day of the year proved to be the second-coldest day in history for many eastern cities, for the daylight sun was capable of nudging the mercury only slightly above zero. New York officials estimated that their coal supply was less than one day from extinction, and federal officials moved swiftly to relieve the burden carried by the nation's largest city. The country's railroad lines, particularly those emanating from such coal-producing cities as Pittsburgh and Erie, were ordered swept clear of all passenger trains to permit a right of way for those carrying precious coal. The Federal Fuel Administration ordered six "lightless" nights a week, saving only Saturday night, when oil- and coal-powered lights could shine undimmed. The Great White Way—New York's Broadway—shuddered in the darkness of the New Year, while courts, banks, and businesses closed their doors with darkness.

As the cold persisted with very little moderation in temperature, major river arteries froze over solidly, shutting off access to coal barges, and solid masses of ice engulfed the boilers of locomotives, hampering further the efforts to restore heat to the civilian and military population of the Northeast. New Year's Day, a major section of Norfolk, Virginia, burned to the ground, and firefighters were frustrated in their effort to save the business section as one hydrant after another froze over. While the average daily temperature in the East hovered near zero, the high plains of Wyoming, Idaho, and Nevada were enjoying their warmest January weather on record. Undoubtedly, the jet stream was contorted rather sharply this winter, reflected in the extreme conditions of east and west.

By the third of January, coal poured through the Pennsylvania Railroad's tubes, hitherto sacred to passenger traffic, and fireboats plowed through the ice floes to release the coal barges. The nation's mines and railroads were now completely in the control of the federal government, and police moved in to confiscate hoarded coal, which they proffered to the poor. By the end of the first week of the New Year, a major snowstorm moved out of the southwestern states, bringing blizzard conditions to much of the beleaguered Midwest. Chicago lay under seven-foot drifts and a howling northwest wind of thirty-five miles an hour. Five days later, the snows began anew, but this time, the entire area from the Rocky Mountains to the Appalachians, from Canada to the Gulf of Mexico, was affected. A new outbreak of polar air followed in the storm's wake, and temperatures plummeted to fourteen below in the Windy City. "No," became kind of a prefix to many items that in ordinary times were taken for granted. No trains, no taxis, no coal, no milk, no cattle, no hogs, no sheep, no schools, no unessential travel, no business after three. Two days later, yet another storm swiped the Midwest, snarling major cities there in snow, perhaps more than ever before or since known.

On the sixteenth of January, the federal government, then headed by President Wilson, made the unprecedented move of shutting down all industry in every state east of the Mississippi, as well as Minnesota and Louisiana. The country was startled by the news, and all but those in the food-producing industry were among the tens of millions made idle in this energy-conserving maneuver. It was unquestionably the most drastic measure ever adopted by this country—in times of war or peace—since its founding. The action was greeted with askance by Congress, but ultimately the president prevailed, and workers remained idle for the following week and for every Monday until the twenty-fifth of March.

A new wave of cold air reached the East on the twentieth of January, as evidenced by a -30°F reading in Clearfield, Pennsylvania, and after that, temperatures returned to more normal levels for the balance of the winter. While the nation's poor suffered in inadequately-heated apartments and tenements, and many soldiers were housed in flimsy barracks with inadequate heating systems, the cruel influenza epidemic was yet to exact its toll. The number of lives claimed in this winter season may never be known, but it is reasonable to conclude that it numbered in the thousands, owing to the invasion of a deadly influenza virus.

Under similar circumstances, America's involvement in World War II was greeted by an outbreak of harsh weather in the waning days of that year. Temperatures dipped to readings well below zero for the better part of a week, and to those who struggled with the depleted stocks in wartime, 1917, the icy blasts were no doubt reminiscent of the tribulations they endured during America's involvement in the first world war. Once again, the cold struck just in time to herald in the winter solstice with another genuine zero day. Owl's Head, New York, recorded an all-time low of -44°F, as did Syracuse with a reading of -26°F. Many thermometers in Upstate New York, designed to measure the temperature only down to -40°F, broke.

Fortunately, this cold wave was not nearly so persistent as the one in 1917-1918, but the irony was that many larger companies led the way in converting from the new wartime fuel—oil—back to coal, which was now plentiful. Near the end of the month, a plan to ration canned goods throughout the nation was effected, but a thaw had set in, enough to send the Ohio River on a wild rampage through Pittsburgh, eleven feet above flood stage. The warmer air overriding the cold surface air caused an outbreak of freezing rain in central Connecticut and loosened about fifty tons of rock and earth on a bus loaded with defense workers, claiming twenty-five lives near Aliquippe, Pennsylvania. Obviously, it's better to have one's cold in moderation.

3.
THE COLDEST MONTH OF THE CENTURY

OCCASIONALLY, DURING THE COURSE OF ORGANIZING THIS BOOK, IT HAS BEEN DIFFICULT TO decide whether a particular era in weather history belonged in the chapter addressing floods or outstanding storms, heat wave or drought, cold wave or snowstorm. Such is the case in the less-than-spectacular winter of 1933-34, when February 1934 was the standout month both in terms of cold and snow. The decision was based upon the fact that the month had twenty-eight days of cold and only a few days of snow, even though—combined—they were worthy of mention in the snowstorm chapter. Such was the case in the winters of 1856-57 and 1857-58. More modern and all-time records for extreme cold fell in February 1934 in the Northeast than in any other month, as we shall see in the succeeding pages.

Actually, the weather took a turn for the peculiar as far back as July 1933. That month was the standout month in terms of heat throughout the East, as temperatures in New York City crested as high as 102°F, within two degrees of the all-time record. September's rainfall of more than 10 inches (in the absence of any tropical disturbances) was three times the normal catch and the greatest since 1882. In one twenty-four-hour period, between September 14 and 15, weathermen in Central Park measured an astounding 5.4 inches. After a brief respite, the weather resumed its extreme ways in November and December, scoring the coldest averages since the harsh conditions of 1917. A very heavy snow blanketed the ground with nearly a foot reported in most eastern cities the day after Christmas, a foreboding of things to come. After a tranquil January, a veritable explosion of winter weather descended upon the United States, from the Rockies eastward.

Russell Owen, a former Antarctic explorer and resident of Connecticut at the time, complained, "No matter how bad it was down in Little America, a man could get out and walk around because the snow had frozen hard. Here, I can't walk 100 yards from the house. We're living in the utmost primitive fashion in the heart of a civilized community. Little America was never this bad." Part of what prompted his remarks was the fact that the snowplows had broken down in the huge February drifts, the electrical power had failed, and the water pumps were idle. This man's opinion of the month of February 1934 is a ringing endorsement of its extreme conditions.

How cold was it? It was just ordinarily cold for the first week of the month, perhaps five to ten degrees below normal for that time of year, when a second, more powerful advective blast from the north put the clamps on the mercury and many all-time records. The mercury plunged to -42°F in Van Buren, Maine, on the seventh of the month, and the icy fingers extended nearly all the way down the East Coast to Palm Beach, Florida. Miami remained in the balmy seventies throughout this period of trial in the Northeast and Mid-Atlantic. Motorists were able to drive safely from Bay Shore to Fire Island (New York) on frozen waters, and Long Island Sound was observed by aircraft to be frozen solid from Stamford, Connecticut, to the Long Island shore for the first time since the winter of 1917-18. By the morning of the ninth, new record minimums, which still stand, were established in Boston and Providence (-18°F), White River, Ontario (-52°F), High Point Park, New Jersey (-47°F), New York City (-14.3°F), Philadelphia (-11°F), and Washington, D.C. at -6°F. In addition, all-time state records fell by the wayside in Massachusetts (-27°F), Rhode Island (-22°F), Pennsylvania, Maryland, Delaware, and Virginia.

Fortunately, between the wars, there was adequate fuel, even for the poor, and people hardly even glanced at their thermostats, while oil gurgled in the basement furnace. It was, in short, the coldest day since January of 1857, and the chill in most southern locations has never been exceeded and, in northern climes, not until the cold winter of 1970. Ironically, February 9 marked the sixty-fourth anniversary of the National Weather Bureau's existence, and they had their work cut out for them. A snowstorm developed over the southeastern states, in response to the extreme variation of the two air masses, and a killing hail and snowstorm varnished the Florida citrus groves as far south as Palm Beach. It was one of the heaviest snowstorms ever to strike Tallahassee and Jacksonville, with sixty-mile-an-hour winds glazing palm trees and citrus groves as far south as Miami. As the storm progressed northeastward, it lost much of its punch, leaving behind only a couple of inches of newly fallen snow but pulled down another arctic blast from the Hudson Bay. This time, the cold formed an ice bridge in the lower Niagara River which joined Queenston and Lewistown and extended well out into Lake Ontario. Many residents along the East Coast noted a freezing mist which sent clouds of vapor steaming well into the air like Old Faithful. Residents of Ocean City, Maryland, were startled to see the ocean "boiling" with this phenomenon. Many people living on the outer islands along coastal waters were effectively marooned from food and medical supplies, as the depth of the ice measured 20 inches thick.

While the weather then moderated for a few days, the contrary features of the weather map were not to be denied. A snowstorm gathered strength along the East Coast and dumped between ten inches (New York City) and twenty-two inches (New Haven, Connecticut) of new snow as the intense low stalled off the Block Island shore with an impressively low barometric reading of 28.73 inches. Strong northeasterly winds drifted this snow into monumental fifteen-foot drifts from the eastern shore of Maryland to the New England coastal waters. As might be expected, several vessels capsized during this storm, and stories of heroic rescues circulated among seamen. To add to the mystique of the month, Northern Lights were vividly displayed on the nights following the snowstorm, suffusing reds and greens and yellows in "lightning flashes," making the scene as bright as twilight upon the newly fallen mantle of snow.

In the wake of this storm, temperatures again descended to zero as far south as Philadelphia, and winds as high as fifty-four miles an hour accentuated the cold. On the twenty-fifth of the month, a new snowstorm traversed the midsection of the country, causing deadly tornadoes in the South, where twenty-five people were killed in Georgia, and new snows in the North. Packing gale force winds, the storm equaled the snow totals of the previous week as far north as Quebec, where drifts of thirty to fifty feet were common. Even the coastline, where snow generally turns to sleet and rain, recorded twelve inches at Atlantic City.

A new cold wave developed in the wake of this storm, insuring a place for February 1934 in the record books. Temperatures averaged generally ten degrees or more colder than normal throughout a wide swath of the eastern United States. Although these records were challenged in the wild weather decade of the 1970s, most of the extreme temperature lows still stand. Occasionally, as in the winters of 1976-1979, the entire winter merits recognition, but the usual force of Nature is a sporadic unpredictable surge of outstanding weather such as occurred in the historical month of February 1934.

4.
THE COLDEST WINTER IN NINETY YEARS

Some of Jimmy Carter's most hectic days in office of the presidency occurred in his last ten days as well as his first. While the Iranian hostage situation came to a head in the dwindling days of his term in office, unpredictable factors in the weather plagued him right from the onset. The problem of diverting natural gas from the ordinary channels to the areas that needed them most took priority over practically everything else in his first days in office. It also set the stage for fuel rationing and renewed discussion of this country's "energy crisis." Faced with factory and school shutdowns forced on account of the unusually persistent cold and lagging fuel production, his administration was immediately taxed as large areas of the country, from Florida to Upstate New York, from Minnesota to Ohio, qualified as federal disaster areas. One of his immediate measures was the symbolic gesture of turning down the thermostat at the White House. Questioned by reporters, he drew more than a few chuckles when he cracked, "I am wearing heavy long underwear; it's cold in the White House." Perhaps the era of reduced federal spending was looming on the horizon.

By the middle of January 1977 most major river systems were effectively closed in the Midwest, namely the Mississippi, the Ohio, and the Illinois. Thus far, winter temperatures had averaged more than five degrees below normal, or better, in nearly every state east of the Rocky Mountains. Meteorologists blamed the chilling turn of events on a blocking ridge of high pressure that built resistance to passing weather systems along the Rocky Mountain states. Consequently, very little moderation or change to the pattern of icy blasts took place in the East. One bitter cold wave followed another, in an endless sequence over the course of the winter. The country's midsection was pounded with repeated onslaughts of bitterly cold air. These conditions translated into one of the century's coldest winters to date, and residents of the eastern two-thirds of the nation knew it was growing worse. Unfortunately, locked in the ice of the nation's river systems were barges of oil, intended for delivery to ease the plight of Chicago, Cincinnati, St. Louis, and other major cities in the Midwest suffering from the brutal cold.

On the seventeenth of January, the cold intensified as Campbell, Minnesota, reported an absurd forty-one below zero, while Chicago shivered at minus nineteen. New snows broke out in the East, but the nation took a temporary break from the weather as front page news as all eyes focused on Utah, where Gary Gilmore received the execution that he had sought for so long. It was the first execution in this country in nearly ten years, and one that caught the public's imagination, for the circumstances of this legal ordeal had been described in the press for nearly three months and resulted in Norman Mailer's Pulitzer Prize-winning book, *The Executioner's Song*. Otherwise, events out west were equally astonishing, for they were in the midst of a particularly balmy winter, with practically no snow reported in the Rocky Mountains, nor any precipitation at all as that section of the country's drought was worsening day by day.

Meanwhile, in the East, schools and businesses were closing their doors indefinitely as the fuel shortage grew critical. Ohio and Pennsylvania were particularly hard hit, as much of their fuel supplies depended upon natural gas, which could not be replenished fast enough to meet the mounting demand. Dayton, Ohio, reported that temperatures remained below zero for fifty-seven consecutive hours. Snow was falling in Ft. Lauderdale and Miami for the first time since weather records had been kept, and the vital orange and vegetable crops of that region were being destroyed. Tens of thousands of migrant farm workers were out of a job. Lake Erie, the shallowest of the Great Lakes, was frozen bank to bank for the first time in modern history. It had been snowing upstate, ever since there was an upstate, and this year was no exception. Temperatures had remained below freezing every day of the winter, and the 150 inches of snow that had piled up since the beginning of the season lay in great white banks, virtually unaltered—only redistributed—since the day it fell. Residents of Buffalo, Rochester, and Syracuse were accustomed to severe winter weather—that area of the country is not known for its moderate weather, but were vexed to find that while their highways and major thoroughfares were perfectly clear of snow, the secondary streets were still clogged.

The deputy mayor of Oswego was asked if the cold and snow made him dread winter, and he replied, "Not at all. I think what I enjoy most about the snow is the community spirit it generates. After a big fall you see people going to their neighbors, asking what they can do to help. Frankly, it's one of the most enjoyable things about living here." Certainly, his enjoyment was not to survive the winter. While the weather may have provided employment to those who predict it, statisticians who measure it, those who write about it, and the salt spreaders and plowers who dispersed it, many a resident could not see the humor in it all. After a follow-up storm on the twenty-third, one man who was obviously disgusted said, "I had a good Cadillac and now I have a bad Cadillac. It is buried under a snowdrift and maybe I won't see it again until the snow melts." Others were not to be so lucky. The National Guard was called upon to do the lion's share of snow removal, and one of their huge personnel carriers rolled over an abandoned car hidden in a drift. "They didn't even know it was there until they felt something give underneath them. It was the roof." It was scary to think of what they might find when the weather settled enough to start digging out their cars in earnest.

Things weren't about to get any better. A new blast of arctic air was surging southward into the country's midsection after leaving the state of Minnesota with devastating wind chills of minus 100 and more. Schools in the fuel-starved regions of Ohio, Indiana, and Pennsylvania were already forced to close, and businesses were closing their doors as their fuel supply was cut off entirely. Blizzard conditions accompanied this front as sixty- to seventy-mile-an-hour winds blasted the new snows horizontally throughout the eastern two-thirds of the country. In the Central Plain states and Midwest, conditions were described as "unmerciful." In Canton, Ohio, where the wind chill stood at minus forty-five, blinding snow pushed across the blacktop roads, "looking like some sort of mysterious white elm from an old horror movie." In Chicago, the wind and cold ripped windows out of high-rise buildings and stranded hundreds on Interstate 65, in Indiana. They were finally rescued by snowmobiles. Much of the midwestern states already resembled ghost towns, as power and light companies had previously cut off all supplies to commercial, industrial, and governmental customers.

One young woman, who was gingerly manning the phones, noted, "Everyone's gone." She continued, with a touch of curious pride: "The chill factor is 80 below and they've cut the heat to 40 degrees." It seemed that many throughout the Midwest were adjusting to the weather with just that sort of aplomb. In West Virginia, on the night of January 28, Governor John Rockefeller took to the emergency broadcasting system as temperatures and visibility plunged to zero. He advised people to stay at home or seek immediate shelter. "Don't travel anywhere. Establish a buddy system." It was not a very pleasant evening. Toledo, Ohio, reported in a dispatch, "All the roads are closed. As soon as the plow goes through, the winds come along and the roads are right back shut." In Buffalo, New York, the plows had long since been abandoned. Nothing but the wind was getting through.

Buffalo is not known for its moderate winter weather, resting on the receiving end of Lake Erie's wintry blasts, but this was the "worst storm ever known," with northwest winds clocked at sixty-eight miles an hour and the thermometer slumped to a frigid minus ten. The city had been virtually paralyzed all month from the accumulation of 150 inches of snow, and now the new storm generated waves of new drifts. Buffalo ran out of places to store their wintry heritage; that hapless city was totally shut down for a week, with no trains, buses or planes coming or going. With no other place to put the snow, they had to physically pick it up and haul it away, once the trains started moving again. Snowplows manned by National Guardsmen fought a losing battle with howling winds and drifting snow, and the dark clouds that hung low over the region cast a pall of gloom that pervaded a landscape of snowdrifts and an arctic-like silence. Most vehicles were only snow covered lumps by the side of the road, and frequently provided landmarks to determine where the road was in the first place.

On the lee side of Lake Ontario, Watertown was literally buried under seventy-one inches of new snows, which drifted twenty-five to the roofs of houses. The blizzard struck there with such fury that thousands of people were caught unaware. The snow piled up at the rate of four inches an hour, and many families were not heard from for another ten days. Food and medical supplies were dropped over the countryside by army helicopter, and in nearby Barnes Corner, a perennial winner of snow contests in the United States, the raging blizzard brought the season's total to 296 inches, enough to ensure that dubious title for them once more.

An amusing article was printed in the Cincinnati *Enquirer*, a city dubbed the "Paris of the Ohio River," but now resembling Moscow. Entitled "Twenty Guaranteed Winter Weather Beaters," the column included, "Volunteer to help old people. Stop wringing your hands. Try needlepointing, crocheting or knitting. Play cards. Pretend you're an American pioneer snowed in for the winter. Do your income tax. Clean the

A scene from Hamburg, New York, on February 7th, following the blizzards of early 1977. (*National Oceanic and Atmospheric Association*)

basement and closets. Snuggle up with a novel. Bundle up and go sled riding. And finally, if you've caught up with your reading, snuggle up with your spouse or a friend. He or she might prove more stimulating than a book."

How cold was it, and why did it happen? Twenty-six cities, including big ones like Chicago and Pittsburgh and southern ones like Jacksonville and Nashville, experienced average temperatures that were the lowest since records were kept. In the month of January, Cincinnati registered a whopping twenty degrees below the expected norms. Nearby Indianapolis was a close second with a minus eighteen. While warmer weather arrived during the second week of February and March and April of 1977 commenced to set all time high readings, the weather was seriously out of whack. After several decades of relatively mild, predictable winter weather—during which modern agriculture developed and the world's population tripled—the earth's weather machine appeared to be entering an erratic period.

Actually, the biting cold of the winter of 1976-77 spread into the country's midsection in September, and that autumn was the coldest of this century, in many areas. After a relatively docile December, the persistent outbreak of extremely cold weather in January dominated the headlines for the better part of five weeks. Meteorologists explained that the unusual cold was caused, in part, by a north-south ridge of high pressure over the Rockies, effectively blocking the warmer air from the Pacific. This air was diverted north to Alaska and beyond, and the extreme northern latitudes of the world experienced record-breaking warmth that winter. Meanwhile, a strong trough of low pressure situated over the eastern portion of the continent repeatedly drew down frigid air from the Arctic. This pattern was repeated all over the world in early 1977, as Iceland and Greenland sweltered with above-freezing

Only one customer managed to burrow through to this store in Adams, New York, on February 7, 1977. (*National Oceanic and Atmospheric Association*)

temperatures and Western Europe shuddered with the eastern two-thirds of the United States. No one knows exactly why this pattern developed to such an extreme; the only speculation available was in the differences of sea-surface temperatures in the Pacific.

Spring was welcomed in the East with more than the usual fervor, and she responded in kind. Temperatures soared to unseasonable levels in the early days of spring, registering highs in the nineties over Easter weekend. It seems that extreme weather begets extreme weather. How easy it is to forget the turmoil that the entire nation was mired in that winter, or to dismiss it as an aberration. Americans are generally unaware of the overall effects of droughts, freezes, floods, and prolonged spells of heat and cold. These are problems generally consigned to other parts of the world and have been perceived and reported as isolated incidents that could not directly affect them. A strong point was made in the winter season of 1976-77.

5.
TEN DAYS THAT REWROTE THE RECORD BOOKS

WHEN NEW YEAR'S DAY DAWNED ACROSS THE CONTINENTAL UNITED STATES, MARKING the beginning of 1982, Miami, Florida, was basking in a steamy eighty-seven-degree air mass while Fairbanks, Alaska, endured a rigid -43°F. This 130-degree discrepancy was an ominous sign of things to come, for the weather machine is often at its worst when such extremes in temperature exist. Things began to happen by the third of the month, when Marin County, surrounding San Francisco, was pounded by more than ten inches of rain in a two-day period by a storm stalled along the Pacific coast. Serious damage resulted from the ensuing floods and mudslides, with houses sliding downhill and major highways blocked. Thirty-seven people were either killed or listed as missing. Further east, a record snowstorm was underway in Wisconsin, and Milwaukee and its suburbs reported between eight and sixteen inches of new snow, the city's worst storm since 1947.

Meanwhile, the frigid reservoir of air overlying Canada was getting even colder, with an astounding reading of -73.3°F reported on the Ross River in the Yukon. Two of the coldest air masses ever to invade the North American continent were soon to follow, slashing records in nearly every locale east of the Continental Divide. The first outbreak overspread the American Plain states on the ninth as Minnesota and Wisconsin reported temperatures of -36°F and -35°F, respectively. By the tenth, the core of the air had filtered into the Midwest, and Chicago (-26°F) and Milwaukee (-25°F) recorded their bitterest daylight in history, amid swirling snowstorms and icy winds. The wind chill in Fargo, North Dakota, nearly fell off the chart with a -98°F reading. The temperature for the Cincinnati Bengal-San Diego Charger game in Cincinnati was -9°F with a wind-chill index of -61°F. George Roberts, the San Diego punter, alleged, "It was kind of like kicking bricks."

Even as far south as Atlanta, where the trial of Wayne Williams was delayed on account of the weather, the temperature dropped off to five below zero, the coldest reading since the spectacular temperature antics of 1899. Boone, North Carolina, reported the mercury hovering at -19°F in a howling seventy-miles-an-hour gale, and in Louisville, Kentucky, the wind chill reached a record depth of -60°F. Lake effect storms dropped a record total of 25.3 inches in a twenty-four-hour period in Buffalo, and when it was all over, that city lay under nearly three feet of freshly deposited snow. Undaunted, the cold invaded the citrus area of Florida on the twelfth, damaging the Florida orange crop and spawning a huge ice- and snowstorm that was to plague the normally immune South for the next two days. Atlanta was in a tizzy under six to eight inches of snow and ice, while joggers in New York ran barelegged through Central Park and some folks swam in the icy waters off Coney Island in defiance of the arctic winds. Meanwhile, in Chicago, temperatures had soared well into the teens, in an impressive warmup.

By the thirteenth, nearly every state in the union reported snow cover, and the storm in the South Atlantic made a fateful turn up the East Coast, spreading gobs of snow to the Mid-Atlantic and New England. In Washington, D.C., where an Air Florida 737 jet bound for Tampa had just been cleared for take-off despite the heavy snow, tragedy was about to strike. With an accumulation of ice on its wings, it just barely cleared the runway, but did not clear the span of the 14th Street bridge just a mile away. First, there was a deafening roar, and then there was silence as the craft skidded across the grey ice and sank in the icy waters. One eyewitness to the crash reported, "I heard it coming. I couldn't see anything. It was snowing. Then I saw the plane coming out of the sky. The nose was up, tail down. It was so loud, I couldn't hear myself scream. And then there was no sound. You couldn't even hear the plane going into the water." Lloyd Creger's tale was relived by many others, for the incident took place about 3:45 P.M., at the height of an early commuter exodus.

Joseph Sticey, a pilot and a passenger on the plane, noticed that the craft did not seem to have enough traction. After it ran out of runway, the plane reared up and immediately stalled. "I looked out," Sticey related, "and I knew the bridges were there. I said to my friend next to me, 'We're not going to make it. We're going in.' "

Once the rescue operations got underway, the ice floes in the river impeded boat traffic, and only five of the seventy-nine aboard managed to grab the life-saving yellow ring from a hovering helicopter, lifted to safety. One victim, who had heroically handed the ring to four fellow passengers before his own rescue, grasped the line for a moment, was half lifted from the water, and lost his grip. Then he slipped back into the frigid waters and vanished from sight. Only moments later, and less than a half mile away, a subway-surface train derailed on the ice and snow, killing three. It was a day eerily reminiscent of the Great Knickerbocker Theater Disaster, detailed in another section of this book.

The following day, a new storm took shape in the South, dumping heavy rains in Florida and again spreading snow northward along the Atlantic Seaboard. Half a million homes in the South were without electricity and heat as another unprecedented load of snow blanketed Atlanta and points north and west. Snow still fell heavily along the lee side of the Great Lakes, and zero-degree weather remained entrenched in the Midwest. On the sixteenth, perhaps the coldest air mass ever to invade this country marched south and eastward, dropping temperatures in Amarillo, Texas, from 55°F to -1°F and to an even greater extreme in Oklahoma City, creating an overnight fall from 62°F to 0°F the following morning.

Cold Sunday, January 17, 1982, is among the five coldest days known in nearly two hundred years of American weather chronicle. The list of cities in which all-time and/or January records fell is as impressive as it is long. A sampling: Buffalo at -15°F, D.C. at five below; Jackson, Mississippi, five below; Philadelphia, minus seven; Nashville, minus eleven; -22°F in Akron; and a stunning -34°F at Eagle's Rock, Maryland. Tower, Minnesota, set an apparent new temperature low for the nation when it checked in with -52°F. A similar reading was reported from Embarrass, Minnesota. Meanwhile, a 136-mile-an-hour wind slammed off from the Rocky Mountains and into the vicinity of Boulder, Colorado, tearing off roofs, ripping up trees, and sending large debris flying into the air. One resident observed that the sky "seemed to be filled with gravel and bushel basket sized wood and debris, nearly one hundred feet off the ground. It was like a meteor shower." Sustained winds of more than 100 miles an hour lashed the High Plains of eastern Colorado for more than ten hours. Winds were to resume exactly one week later in the same region, and with the same force. Known as a chinook in the West, these winds frequent the Mountain states in the winter and spring and frequently bring unprecedented temperature rises. This particular wind warmed Boulder from 21°F to 58°F in a matter of six hours. For most residents east of the Rockies, January seventeenth was a day to remember, for temperatures in most eastern sections failed to rise above zero.

While the cold abated somewhat after the eighteenth, the snow did not. Minneapolis, Minnesota, was soundly lashed by their most intense snowstorm in history, a whopping 16.6 inches in only thirteen hours. Another intense winter storm formed in the Southwest only a day later, and pushing another heavy band of snow northward, and the Twin Cities established still new records for intensity and total snowfall—another twenty inches this time. Blizzard conditions prevailed throughout the Great Plains, and an avalanche left 3,000 skiers snowbound in Utah. The High Sierras in California and Nevada received record total snowfalls throughout the winter, snow that was still piled five feet thick in mid-June.

Satellite photos showed that more than three-quarters of the land surface of the country was snowcovered at this time, more than in any other January of record. The ocean itself froze for several hundreds yards from shore, off the relatively mild waters in Atlantic City. Near the end of the month, another twenty-two inches of snow descended over the Midwest and St. Louis recorded its largest snowfall (19 inches) since 1890.

The bitter cold of 1982 lasted through the middle of February, when Indianapolis, Indiana, reported a minimum of -21°F, the coldest February weather since the Great Arctic Outbreak of 1899. But by the thirteenth, Bismarck, North Dakota, recorded its forty-fifth consecutive day below zero. While most winters present a sampling of one extreme or another, the middle ten days of January 1982 must be recorded as the epitome of extreme cold due to the extent and nature of the myriad records which fell. Ironically, winter was not through yet. In the early days of April, a major northeaster with blizzard dimension stalked the East Coast, dropping between six and thirty-six inches of new snow over a wide area. Only three days later, after record-setting cold in the East, a second storm dropped additional snow along the same seaboard route.

II. SNOWSTORMS/BLIZZARDS

No word, meteorologically speaking, inspires so much dread and delight among people as the word snow. *People talk about the weather every day—a sort of social welfare, one in which nearly everyone can venture an opinion—but mention snow and a response is virtually guaranteed. One always remembers the deep snows of their youths, of white Christmases past and the seeming lack of snow in their adult lives. Historically, there was no preference for deep snows in the nineteenth century as opposed to the twentieth, nor were any deeper snows observed earlier this century than those of the past few years. We remember all the big ones, as though they occurred in one season, and that was the way things were, season after season. Nothing could be further from the truth. In fact, being younger, being mired in a knee-deep snow was easier back then, for our knees were much closer to the ground.*

Snowstorms have occurred at virtually every latitude in the United States over the years, excepting extreme southern Florida and the Florida Keys, the former location having observed traces of snow from time to time. There is preference, however, for snowstorms to develop and intensify in the area from the Rocky Mountains eastward. Winter storms generally originate in one of four places in the country. First, there is the area just east of the Rockies, in Colorado south to the Texas-Oklahoma panhandle. These storms generally track eastward, or northeastward to the mid-Atlantic and New England states. Occasionally, they spawn a secondary low off the Carolina coastlines and move up the coast in the form of northeasters. Secondly, the Gulf of Mexico is responsible for heavy winter weather, east of the Mississippi, for it becomes an active storm breeder in November, and the peak of that activity crests in the winter months of December, January, and February. Warm, moisture-laden air is in frequent conflict with polar air from the north, and storms tend to form along the polar front in winter. Usually, these storms from the gulf will run up the coast in eight to twelve hours, but in the case of the Blizzard of 1888 and many other coastal disturbances over the years, it was blocked by a strong high-pressure cell over New England, which prolonged its attendant precipitation and winds.

Similarly, coastal depressions that form off the waters of Flor-

ida and Georgia make a run up the coast, many times paralleling the coast, ensuring strong northeast winds at the surface to keep temperatures low enough for snow while the broad southeasterly component of winds aloft feeds the storm with the necessary moisture for large accumulations. Occasionally, winter storms are bred in this manner off the North Carolina coast, as a result of which the warm moist waters off Cape Hatteras come in conflict with dry cold air from the continent. This type of coastal snowstorm is responsible for huge snowfall amounts when blocked by a Hudson Bay high pressure system in its usual position off the New England coast.

Finally, there is the lake-effect snowstorm, or more accurately, blizzard. Pristine arctic air moves southeastward over the Great Lakes, lifting the considerable moisture available to great heights, and the result is blowing and drifting snow downwind from the Great Lakes region. In late January and early February of 1977, while the West was basking in one of the warmest winters ever known, cities like Erie, Pennsylvania, and Buffalo, New York, ran out of places to put nearly fifteen feet of snow and hauled it out of the region by freight train before it melted and caused floods of Old Testament proportions. Frequently, such precipitation survived to blanket most of Upstate New York and Ohio, Pennsylvania, and western New England, but rarely does the precipitation hold together once it crosses the Appalachian Mountains. Living outside Philadelphia, I was astounded to see such a blast survive on January 28, 1977. The evening had been warm and muggy for that time of year, with temperatures in the middle forties, when a strong wind blasted through the Delaware Valley, accompanied by a violent thunderstorm. Over the course of the next few minutes, rain changed to snow, and over the next two hours the temperature commenced to decline thirty-four degrees and for that period we had a taste of lake-effect snows that so frequently stalk the Great Lakes.

This chapter is intentionally divided into two sections, one covering outstanding snowstorms and the other dealing with blizzards. The distinction between the two is simply that, in order for a snowstorm to qualify as an eastern blizzard, *there must be a sufficiently great quantity of blowing and drifting snow to limit visibility to a matter of yards. Criteria used by the National Weather Service include gale force winds of thirty-five miles an hour or more and temperatures below twenty degrees. An extreme blizzard must occur with temperatures of ten degrees or less and with winds of forty-five miles an hour or more. Ironically, no additional snow is required of a blizzard, just blowing and drifting of* any *snow, including that which already lies upon the ground. Ordinarily, however, substantial precipitation accompanies a blizzard in the East, and the episodes described later in this chapter focus on some of the more extreme cases. A true western blizzard, with temperatures at zero or below, rarely if ever occurs in the East. One further distinction should be made: Conventional wisdom among people with short memories dictates that it can be "too cold to snow." While major snowstorms in the East usually occur when temperatures are at or slightly below the freezing point, when the available moisture supply of the atmosphere will be at its greatest, meteorological history is full of examples of major snow that took place when the temperature was ten degrees or less. One need only read as far as the blizzards of 1888 or 1899 to see the obvious discrepancies in this statement. Ask anyone who wintered over in Buffalo in the season of 1976-77 or who recalls the 1½- to 3-foot snows in the East that occurred on February 19, 1979, while temperatures hovered in the teens, whether it can be too cold to snow.*

Substantial snow ordinarily holds off until the winter months, including the period from December through March, but on October 10, 1979, anywhere from two to eight inches fell over extreme southern Pennsylvania northward to the New York border. Snow has been noted atop Mt. Washington in New Hampshire as early as the last week of August, and many times in the surrounding countryside in September and October. Similarly, late snows have been noted as far south as Washington, D.C., in May, and in the year without a summer, 1816, in every month but July frozen precipitation spread its canopy to the Delaware Bay. On the twenty-seventh of April, 1857, a coastal storm dropped so much snow on southern New England that several buildings collapsed in Connecticut from the weight of the snow. Four years later, on the fifth of May, between one and two feet of snow surprised residents of the mid-Atlantic and New England states. Snow squalls in early June have been documented over the years for a similar area. Nature does not always respect her seasonal boundaries.

Particularly snowy seasons have plagued residents east of the Mississippi River as recently as 1947-48, 1960-61, 1966-67, 1970-71, and in the back-to-back atrocities that befell the Northeast in 1976-77 and 1977-78. Each area of the country has its landmark snow season, and variations in a given locality make it hard to crown the ultimate snow-king. Depths within a thirty-mile radius may vary by a foot or more due to differences in altitude, blowing and drifting, subtle temperatures differences or just plain inaccurate measurement techniques. Suffice it to say that some seasons are particularly harsh, but what happens one winter has no bearing on the following cold-weather season. For example, an uneventful season in 1918-19 was followed by a wild snow season in 1919-20, while an uneventful 1931-32 was followed by an equally unimpressive season in 1932-33. February 1934, on the other hand, was long regarded as the title holder in many parts of the country for extremes of cold and snowfall. It was not to be outdone until the harsh winters of the late nineteen seventies.

Differences in the prevailing wind flow of the upper atmosphere often dictate where the heaviest snow will fall in any given season. In the early nineteen eighties, for example, the snow capital of the country was not Chicago or Buffalo, not the high slopes of the Green and White mountains of New England, but the coastal plains of the mid-Atlantic from Elizabeth City, North Carolina, to Virginia Beach, Virginia. Those years, many an intense storm threatened the Northeast only to dump their respectable loads of two or three feet on the unsuspecting hamlets along the shore, unaccustomed to any snow at all. That period was the height of the drought that plagued the East in the early eighties, and many storms of all seasons failed to penetrate the northern latitudes beyond the lighthouse at Cape Hatteras. Patterns of this sort often develop early in the season, and early clues as to the coming winter weather may be found in the prevailing storm tracks of November and December.

Snow. So pristine, so imaginative, so subtle, soft, and serene, falling at first in such silence as to betray its presence, growing on rooftops like a white moss. Laying on the ground or gently embracing the tree trunks or fence posts in the country, it is so unusually beautiful as to set the child in us free. Lying on the streets and highways, it can be a menace to transportation, and on sidewalks and driveways a peril to the life of the shoveler from overexertion. Most deaths following a snowstorm are attributed to heart attacks from shoveling a weighty snowfall. Brilliant, dazzling, frightening, from another land, snow is bound to visit in varying amounts in each winter season and to depart in the early days of the following spring, like a pathetic yet brave snowman in the brilliant sunlight. The best lesson is to learn to cope, for as history demonstrates, winter's offerings are so often overwhelming and the best course is sit by the hearth, enjoy its beauty, and stoke the fire with the knowledge that it will pass.

6.
THE BLIZZARD OF 1888

NO STORM THAT HAS EVER VISITED THE EAST HAS BUILT SUCH A LEGEND FOR ITSELF, NOR ANY deserve to be remembered as well, as the Blizzard of 1888. Nearly 100 years have elapsed since the waning days of winter in March of 1888, long enough so that virtually no one living today could give a first-hand account of the storm, and yet it stands as the epitome of winter storms and probably always will. Each passing season brings unusually heavy snows to some portion of the country, and the cold winds still sweep across the arctic tundra, plunging temperatures to new lows, but rarely do the forces of snow, wind, and penetrating cold combine to produce such results over so large an area as the famed Blizzard of 1888.

The meteorological circumstances of the storm are in themselves quite interesting. Springlike weather had asserted itself throughout much of the East, sending temperatures well into the forties and fifties over several days preceding the storm. In fact, forecasters had predicted gentle southeast winds with rain arriving only a day before the event that most contemporary meteorologists would rather forget. A storm was crossing the Great Lakes, dragging with it bitterly cold weather on its western flank. Meanwhile, another disturbance was taking shape in the southeastern states, destined, the prophets felt, to bring rain due to the mild temperatures that the East was experiencing. With twenty-twenty hindsight, it is easy to illustrate how they underestimated the impact of the cold front that was charging through the Midwest. That cold air interacted with the developing storm in the South, energizing that system. The consequences were disastrous.

As the storm moved slowly northward, light rain broke out in the middle Atlantic states, and the rain grew heavier as the temperatures in D.C. fell from the fifties to the thirties. Gradually, snow and hail began to mix into the precipitation pattern, and telegraph and telephone wires began to snap. At sundown, on Sunday, March 11, temperatures in the Mid-Atlantic skidded to freezing and the snow began to accumulate. Further north, New York City was experiencing torrential rains, which fell in sheets flooding the streets, as temperatures eased back from the low forties. By midnight, the precipitation there, too, was changing to snow, but few were aware of the incredible sight they were to face the following morning. They might just as well have stayed in bed. Overnight, temperatures slipped through the twenties and the wind shifted to the northwest, howling through the chimneys of the sleeping towns. Later the next day, in the raging storm, conditions worsened while the temperature plummeted to only five above. One can scarcely imagine the additional accumulation of snow had all the precipitation of the previous day been frozen.

There was an interesting account in the New York *Times* about a man who had gone to visit a friend across town on a Sunday afternoon in New Jersey. With light rain falling and the temperature still in the forties, he set out in a light fall jacket, prepared only with an umbrella. As evening approached he could see that things were deteriorating outside, the rain being hurled against the window panes, but that "conveyed no more definite idea to me that it was a particularly disagreeable night outside and a particularly agreeable one inside." By 10:30 P.M. he set off for home but decided better of it, for there was a thoroughly soaking, bone-chilling rain in progress. He was invited to spend the night, and before he bedded down he opened the window slightly. At 4 A.M., he awoke chilled, for he was sleeping under a blanket of snow that had gushed into the window. Cleaning things up with a shovel from the coal bucket, he saw that the snow outside had already drifted to more than two feet. After another few hours' sleep, he arose "to more snow within a hundred yards to keep a polar bear happy for life." The snow was more than three feet on a level at that time. After spending the better part of the day watching it snow, he set out for a train to take him to New York for an appointment at six in the evening. His account follows:

> The snow was everywhere. Great piles of it rose up like Arctic graves ...in all directions. Every way that I turned I was confronted with those awful mounds. I took my bearings and steered for the Jackson Avenue Station. Every step I took, I went into my knees in snow and every other step I fell over onto my face and tried to see how much of the stuff I could swallow. The wind was at my back and its accompanying snowflakes cut the back of my head and my ears like a million icy lashes....
>
> I plowed my way, jumping, falling and crawling over the drifts, some of which were nine or ten feet high, and regaining my wind in spots where the snow had been driven away, and after an hour and ten minutes I got at the end of my six blocks. There were trains there, two of them but they were stuck as fast as if somebody had suddenly dumped hogsheads of mucilage about them. There were passengers in them too, half-frozen and wholly disgusted, some of them women.
>
> I gave up the idea of going to New York...my trip back to the home was simply awful. The wind was straight in my face and beat so in my eyes that I couldn't see a rod in front of me. My mustache was frozen stiff, and over my eyebrows were cakes of frozen snow. My gloves were of kid and by the time I had gotten half way back, I thought my hands would break off at the wrist. I stumbled along, falling down at almost every step, burying myself in the snow when I fell, struggling frantically up only to sink deep down again. Then I began to feel like a crazy man. Every time I fell down, I shouted and cursed and beat the snow with my fists. I was out there all alone, and I knew it, and if I should get down some time and not be able to get up I knew I might just as well say my prayers. Then it got dark. The wind howled and tore along, hurling the icy flakes in my face, and the very snow on the ground seemed to rise up and fling itself upon me.
>
> In one of my crazy efforts to forge ahead, I caught just a glimpse of the welcome gate posts, and then I laid down on my back and hollered. I felt as if I couldn't move a limb if $40,000,000 was held above me. Somebody heard my cries, and just as I was going off comfortably to sleep my friend came plowing out through the snow, and he and his man dragged me into the house. When I woke, I felt like an incipient blizzard myself.

Transportation and communications, of course, came to a complete halt. By

The Blizzard of 1888 in progress. (*Library of Congress*)

the early morning hours of the twelfth, wires snapped from the weight of the snow, and sleet and telegraph poles were scattered asunder in every direction by the wind. In an age when most everyone went to work no matter what the elements, there were some amusing stories to be told about their efforts. One man asked, "Cabby, are you engaged?" "No," was the reply, "and I don't want to be." "I'll give you $5 to carry me to City Hall." "Not for $25," came the answer. The traveler wanted to argue but the cabman whipped his horses into a walk and drove away shouting, "I'm in luck if I can get my team to the stable; I don't know but I'm in luck to be alive." Shortly the horses were led to the stable, or in that general direction.

A woman was seen wading through a gigantic drift, struggling every step of the way, and only halfway through, seemingly overcome with exhaustion, she fell flat in the drift. About half a dozen men who had witnessed her plight came to her aid, and before she could be reached, the fierce wind had completely covered her with snow. The body of a twelve-year-old newsboy was discovered three days later in the snow near Fulton Avenue in New York, frozen to death. While he was not immediately identified, his name was presumed to be Fischer, for he had not been heard from since leaving home on Monday.

Many others, fortunately, found the elements too overwhelming and returned home to spend the day with their families by the fire. The New York *Times* observed how difficult it must have been for "good steady church-going heads of families when they had to get through breakfast without their favorite newspaper, their hot buttered rolls and their fragrant coffee enriched with boiling milk." They began to question "whether life was worth living at all, with all these trials and tribulations to undergo."

Perhaps hardest hit were the trains of the day, and transportation in general. By 11 A.M. on that fateful Monday, every horsecar and elevated railroad train had ceased to run, and the result was a general expression of opinion that an immediate and radical improvement in public transportation was imperative. It seems the Blizzard of '88 accomplished what months if not years of debate and agitation had failed to do. The *Times* noted, "It is due to the elevated roads to say that they ran longer than the street railroads and it is due to the street railroads to say that they did better than cable roads which is not saying much, for they did not run at all."

Most trains did make an effort to run, and everyone without exception became bogged down in drifts, some thirty to forty feet high. While there were some hardships encountered, such as a shortage of food and water, most were supplied by local farmers—at something approaching exorbitant rates.

People marooned on trains going nowhere took to cardplaying, euchre, singing and dancing, and a feeling of merriment prevailed. In most cars, there was whiskey available, and every car was a smoker. For thirty-six to forty-eight hours, until the plows arrived, the men carried on, while what few women there were lavished in the men's attention until they were overwhelmed when they huddled in the rear of the car. One long specimen of humanity muttered, "Well I'll be darned to think that in this 'ere nineteenth century, with the city of New York spreading its confines about on every hand, a train full of people should sit in the depot here and they have nothing better to do than spend the night and day playing euchre and smoking outrageous tobacco."

Some of the younger ones did not. One proposed a twelve-mile hike back to civilization on Tuesday, after spending the better part of a day stranded in the drifts. He was ridiculed, but finally found a party of six to join him in his pilgrimage. The party set out in the New Jersey Meadowlands headed for Carlstadt, twelve miles away, with the wind howling at thirty miles an hour and temperatures about five above. The windchill was in the vicinity of minus forty. There were, of course, the same obstacles confronting them as there were everywhere else—train-boggling snowdrifts. Many they went around, prolonging the trip which lasted nearly ten hours. Finally, as they could see the lights of their destination, they became caught in a huge drift, unable to move, and a wave of panic took possession of them. The nearest house was only 400 feet away, so they lunged forward, crawling on their hands and knees as the crust of the snow would better bear their weight. This last unsurmountable distance took nearly an hour, and when they were rescued they were taken home to thaw out by the fire and tell their stories to their awe-struck families.

By Wednesday afternoon relief was on its way, and a sensation spread over the desolate souls trapped in the trains. A snowplow came up the road throwing the snow high up into the air, undoubtedly affording a beautiful sight. All hands rushed out to see it and went up close to the tracks. When the train passed in a cloud of white, it "was laughable to see the passengers digging themselves out of the snow which had been thrown over them." The next plow that went through was viewed from a more respectful distance.

Tuesday morning dawned in an unprecedented spectacle of a snowfall that had lain untouched for the better part of thirty hours. Snow was piled, in some cases, to the second-storey win-

Proud veterans of the blizzard pose before heaps of snow. (*Library of Congress*)

dows, and people gradually began to tunnel out of their homes. It was not until Wednesday that the novelty wore off and most decided that it was time to do something about it. When the sidewalks were cleared, piles of snow along them were built up so high that one could not see across the streets. Many men built caves in the big banks and built fires in them to melt the snow, or turned streams of hot water upon the big heaps. Snow began anew Wednesday afternoon, and it was forecast to be a heavy fall, but it did not materialize. The snow changed to rain, and by sundown the sun peeked through the clouds. Overall, the situation was accepted with universal good nature, and one pile of snow, about fifteen feet high, bore a sign reading "This snow for sale." There were no takers.

In retrospect, it is most fortunate that this blizzard occurred in the middle of March rather than January, when it would have taken weeks—if not months—to dissipate. There were hundreds of ships lost at sea, and the death toll numbers in the hundreds among mariners. Particularly hard hit were the Maryland and New Jersey coasts. Every state had a taller tale to tell, and even as far north as Ontario, where "only" fifteen inches of snow accumulated, winds were measured at seventy miles an hour. How much snow did fall? It's difficult to answer that one, but on an average it was two to three feet in the mid-Atlantic states, with as much as five feet reported in Maryland. But the fact remains, the storm had so much universal impact that even the "oldest inhabitants" could not recall its equal. One man remarked to a European visitor, who had never seen the likes of this storm, "This is weather. What we usually have is climate."

7.
THE COLD SEASON OF 1899

It seems that so many of the snowstorms discussed in this book occurred on a Monday as to render the phenomenon uncanny; yet the blizzard conditions which prevailed throughout the nation on thirteenth of February, 1899, were no exception. Temperatures over the course of the first two weeks of that month ranged fifteen to twenty degrees below normal throughout most of the United States, and the fierce winter winds created a penetrating windchill unheard of at that time—or any other, for that matter. The Gulf of Mexico was quite active in that season, spawning storm after storm which coated the East with a foot or more of snow which lay on the ground without dissolution due to the subfreezing temperatures. After the storm passed on the eighth of the month, it set the stage for some of the coldest temperatures ever known east of the Rockies.

Temperatures in the Mid-Atlantic region skidded from thirty-two degrees to five below zero overnight, accompanied by an unrelenting fifty-eight mile-an-hour wind from the northwest. While it seemed that the cold equaled the conditions of the Blizzard of 1888, no one ever compared cold spells or any other weather conditions with the extraordinarily freak elements of that year. Had anyone bothered to research, they would have found that the extraordinary winds and temperatures of 1899 exceeded those of 1888. But, according to the New York *Times*, "many persons have the utmost confidence in the fallibility of the official weather prophet and there were pessimistic mortals who believed that the day would bring bright spring weather."

Tragically, it did not. Fuel was scarce and it was invariably the poor who suffered the most. Perhaps the most graphic example of this was the story of a four-year-old child who lived with his mother in a tenement in New York. One morning the child, Allison Munroe, awoke and climbed out of his crib to view his mother lying on the cold living room floor:

> "Mamma," he said, "are you asleep?"
>
> The hard set lips of the woman on the floor made no movement in reply.
>
> "Mamma," the boy repeated, dropping down beside the still form, and rubbing his chubby hands over the cold face, "wake up. Allison wants his breakfast."
>
> There was still no movement of the pale drawn lips, not a flutter of the closed eyelids. The baby pressed his own warm lips to his mother's cold ones. His kiss had never failed to wake her up when she slept, and he wondered that she failed to open her eyes then.
>
> "Mamma must be tired; Allison wait til she wakes up," he lisped half to himself.

Boston was heavily hit by the snowstorm of 1899, the worst in that city since the blizzard of 1888. Shovelers are seen working hard to clear the railway tracks. (*New York Public Library*)

> And so he waited, but his waiting was in vain for his mother's sleep was that of death. The hours dragged on by and night came, but still the boy waited, chilled to the bone and weak with hunger. At last, he fell asleep, his arms about the form so strangely still, his face pressed lovingly against the one so white and cold.
>
> Wednesday came and went, but still the boy crying bitterly and appealing repeatedly...kept faithful watch beside the silent form in the bleak room. He felt his little limbs grow numb, and found that he could not stand upon his feet. But still, he clung to the dear form so strangely still to him. Finally, there was a knock on the door....

The boy, amazingly enough, survived, but the fate of his mother was the same one met by scores of others during this cruelly cold period, particularly among the indigent.

How cold was it? A sampling of temperatures on the morning of the eleventh are revealing: Washington, D.C., minus eleven; New York City, minus eight; Pittsburgh, minus twenty; Tampa, Florida, eighteen degrees; Galveston, eight degrees; New Orleans, twelve degrees; Tobyhanna, Pennsylvania, minus seventeen degrees. Overnight, temperatures deliberately reached for new records, plunging as low as minus fifteen in D.C. on the twelfth, a record which still stands. Cattlemen reported that the ears of their hogs fell off from freezing, the least touch causing them to fall to the ground. Overnight, a storm was taking shape off central Florida, destined to move up the coast spreading snow and new discomforts in its wake.

Actually, it was snowing nearly everywhere in the nation as dawn broke on the twelfth of February, and the weathermen, without the aid of radar, computer printouts or satellite photographs, were at a loss to explain where it was all coming from. The latest thing in blizzards swept east on the evening of the eleventh, and the Saturday night theatergoers were greeted with blinding white sheets of snow. By Sunday the residents of the East, being inured to being frozen, awoke ready for anything, so being greeted by a blinding snowstorm seemed to be positive relief from the previous days' cold. Sunday was described as a relatively pleasant day, for the temperature rose to dizzying new heights of ten above zero and it became possible for "muffled pedestrians stumbling along the streets to negotiate their way without immediate danger of freezing to death."

Actually, the next few days were about as cheerless and miserable a stretch of time as could be imagined, for the snow was not the comfortable everyday variety but swirled along in minute particles like crushed sand, prepared to bite and sting and find its way down a man's coat collar and into his ears. According to forecasters, there were two or three storms due in the Mid-Atlantic from different quarters, and the resident meteorologist in New York, a man named Emery, passed out the cheerful information that "although this one got here first, it may not be the worst." The whole country was in such a frozen condition that a full-fledged blizzard was likely to be hatched out "at any time, from any quarter." He was certainly right about that.

By Monday, there was no greater chronicle of adverse weather conditions available all the way from Maine to Georgia, from Iowa to Texas. The latest storm to affect the East came from the South, after benumbing and paralyzing that section of the country with a grip of ice such as never before was known there. It left death and destruction in its wake and was met in New

Those who thought they had seen it all in '88 are seen here eleven years later posing at 7th and Pennsylvania Avenue, in Washington, D.C. (*National Archives*)

York by welcoming northwest winds prepared to distribute its luggage to the best advantage, and the frozen ground, already burdened with its ice and snow, was the receiver of an additional blinding white downpour. Needless to say, all of the Lincoln's Birthday festivities were canceled, "even in those sections of the city where pleasure usually halts for nothing." The suburbs were nothing more than white rows with blocks of houses, seemingly deserted, over and around which the wind shrieked and howled all to itself and whipped up great snowdrifts with careless abandon.

By the fourteenth of the month, the storm passed northward into eastern Canada, and into history. The South, Midwest, Mid-Atlantic, and New England had sustained winds ranging between forty and sixty miles an hour. In retrospect, this period in 1899 was more brutal than the week-long carnival posed by the Blizzard of 1888. While an average of only fifteen inches of snow fell throughout the region in the three-day period, temperature readings were easily ten degrees colder, and the wind raged at speeds nearly twice those of the famous storm of 1888. The snow, however, was lighter and fell over a longer period than 1888, accumulating in less dramatic fashion.

Each episode had its freakish elements, but one of the most peculiar in 1899 was the fact that the nation's lowest temperatures were recorded in the Mid-South, with an unbelievable minus thirty-nine being recorded in Lebanon, Kentucky, on the thirteenth. Overnight, earthquake shocks were felt in Ohio, Tennessee, and North Carolina, and by Wednesday morning, conditions improved dramatically as the better part of the nation was released from the fetters of the storm seemingly unequaled in extent and severity. Knoxville, Tennessee, had never before or since held the honors as coldest place in the nation, but when the sun poked its welcome head over the eastern horizon on the fifteenth, the mercury still hovered as low as minus seven with Alpena, Michigan, registering a poor second. The Blizzard of 1899 repeated the lesson that blizzards are not always conceived in the great weather hatcheries of the American Northwest but in the mideast suburban backyards where, on other years on the same date, crocuses and hyacinths have bloomed. Nature knows no boundaries.

8.
THE KNICKERBOCKER THEATER DISASTER

One of the boasts of the management of the Knickerbocker Theater in Washington, D.C., was that hardly a week went by without at least one cabinet member or high-ranking official attending a performance in the fashionable section of the Capitol's northwest section. The movie theater, located at 18th Street and Columbia Road, was one of the largest motion picture theaters in the country and on a good night would pack in 2,000 patrons, who were entertained by a live orchestra as well as the latest in contemporary theater. Inclement weather, to put it mildly, kept most away on the night of January 28, 1922, and the crowd was estimated at 500. Most huddled toward the front sections of the theater, which normally would have commanded a higher price, but it was the theater's policy on such occasions to let people sit wherever they pleased. This was one Saturday night that the best seats in the house were towards the rear.

A powerful storm had formed over the Florida peninsula during the day on Thursday, and by Friday afternoon snow had moved into the Mid-Atlantic states. The disturbance was a slow mover, blocked by a high pressure cell in the New England states. Snow did not arrive in Philadelphia until the following morning, and in New York around noon. Upstate, in Albany, skies were clear, and they received no snow at all. Throughout that evening and all day on the twenty-eighth snow piled up to an incredible twenty-nine inches in Washington, D.C., halting all transit and plunging the city into an arctic ruin. Compounding the problem, temperatures were at or near freezing, and the snow had an unusually high water content, estimated to be between four to five inches melted equivalent. During the day of the twenty-eighth, officials of the Knickerbocker Theater discussed the feasibility of removing the snow from the roof, but they concluded that there really wasn't any danger to the structure.

The tremendous burden of the snow was to prove them wrong. Shortly after 9 P.M., the wind shifted to the northwest, at near gale force, and the storm took on a semblance of a blizzard, driving snow in blinding sheets. One of the theatergoers reported that he heard a hissing sound and had seen the first powdery handful of snow sifting down over the head of the orchestra leader in time to make an escape. From his seat well forward of the balcony, he raced to the exit, and a great blast of air literally blew him out of the theater as the roof caved in. Another one of the injured reported, "The first I knew of the accident was when I heard a scratching noise overhead, and looking up, I saw

Inside the Knickerbocker Theater following the collapse of the roof. (*Library of Congress*)

the roof falling and I tried to dodge it, I was in the gallery and went down with it. Next thing I knew, I found the gallery on the street level. I looked around for my wife. All I could see of her was her hand sticking out from the cement and snow. A chunk of cement had fallen on her right leg." His wife, though badly injured, managed to escape with her life.

As the roof fell in, it took the entire balcony with it, twisting the steel beams into the audiences below. Most of those beneath the balcony perished. Those seated towards the front had little chance except as fate determined that they should be missed by the great girders. They must have realized this, for the arms of many of those whose bodies were carried from the ruins were upraised as if to ward off a blow. Practically everyone in the theater was menaced, and those who escaped with their lives did so by chance.

There was a scene of wild confusion, at first, at the disaster site. Those who first made their way to the auditorium doors saw before them a dim mysterious heap of wreckage faintly lighted by a string of colored lights on stage that still stood in tact and, from the reflected glow from leaden skies above, the void where the roof had been. One of the biggest problems that rescuers faced was untangling the victims from the wreckage that lay all about, piled carelessly above them. Lieutenant V.M. Parsons, called upon from the local Marine Corps, related,

> We were digging into the ruins when we saw a tuft of red hair protruding. We uncovered a small boy, probably nine years old, who told us that his sister was beneath the pile of debris also. The girl was rescued and neither she nor the boy was seriously hurt through some miracle. But their mother, nearby, was dead.

Oddly enough, sleeping peacefully beneath the debris of the wrecked theater were two little girls about four and six who were found nearly ten hours after the playhouse roof caved in. Neither was hurt.

Others were not so fortunate. William Crocker, twenty-six, was sitting next to his bride when the avalanche of broken plaster, bricks, snow, splintered wood, and twisted steel beams catapulted upon the audience, killing him instantly. His wife, from her third-row seat, clasped his hand and called to him as the roof fell, but she only heard a groan and there was no responsive clasp from his dying fingers. Representative John Southwick of Florida was attending the performance and gave a vivid picture of what it was like:

> Suddenly there was a sharp crack. I looked up and saw a great fissure running across the ceiling. It was right over my head. I instantly realized what was happening...while I was looking up, a great piece right over my head started to fall. I ducked, crouching involuntarily, I suppose, between the seats. A piece of the roof struck the seat right where I had been sitting...and pinned me down where I had been crouching. The noise was simply awful. I can't describe it. It was a great tremendous roar. I can never forget it.
>
> In the midst of the roaring were shrieks and cries of women and children and a few shouts of men. There were cries for help, groans and worst of all, the moans of those in terrible pain. I can't describe it. I see it all the time...those poor children and men and women crying and groaning there. I guess there was a lapse of 20 seconds before the balcony fell. Funny, but it spun around, kind of twisted as its supports gave way and it swung down on those below. It didn't go straight down, just kind of slid sideways....
>
> Across the aisle from me when the crash came was a little fellow. I never saw him again and I wonder if he's dead. I crawled over the broken seats and plaster and snow to the door. On the way out, I saw a young fellow half curled up and crying for help. I leaned over to lift him and then everything went black. Next thing I remember, I was at the door, wiping the blood from my mouth and eyes. I don't know how I got out.

The rescue was carried out with acetylene torches, and hacksaws to cut through the iron girders. Members of every police and fire department in the Capital as well as the nearby military installations responded double-time through the knee-deep snow, yet at 2:30 A.M., nearly five and one-half hours after the collapse, the rescue force had not dug down through the mass of fallen cement and steel to the seats on the first floor underneath the fallen balcony. Most of the victims taken out were barely conscious or dead. Several had arms or legs missing and there were many amputations performed on the scene to extricate those trapped. A slow stream of victims trickled out of the theater, to be taken to local hospitals or morgues in ambulances, taxis or private automobiles, and blankets and pillows were made available from neighboring houses. Needless to say, the rescue effort was hindered by roads so clogged in snow that two vehicles could not pass without slamming into a snowdrift. The work of mercy went on despite the continuing menace of the wind swaying the remaining walls, threatening their imminent collapse, trapping the rescuers.

There were grim stories of entire families wiped out, caught beneath the enveloping blanket formed by the falling roof, and each hour brought some new development that added to the horror of the tragedy, as the weary rescue parties labored throughout the night and the following day. A total of ninety-five lives were claimed that night, among them Albert G. Buhler, thirty-two, who died a few minutes after being taken from the wreckage. Witnesses said that when rescuers initially found him, he assured them that he was all right and urged them to take others nearby, including his wife, who lived.

A pitiful stream of mangled bodies, dead and alive, flowed out of the theater for thirty-six hours, and the dead were taken to a makeshift morgue nearby. The dead lay in long double rows. A tearful relative or friend, husband or wife, father or mother would walk by and recognize the crushed form at last. Up and down these aisles of the dead walked those whose fears had drawn them there because of someone missing in the family circle. Women, already weeping in certainty of what they must find sooner or later beneath the kindly blankets, made the journey of sorrow many times before they found what they sought. Men, with worried faces, leaned to draw back the coverings and then gasped with short-lived relief as they moved on to the next huddled form. Many times, both husband and wife died as they sat side by side, but in other instances, only one perished and the survivor had to make the terrible pilgrimage of recognition in the chamber of the dead.

So often the influences of a storm of this dimension are subtle, affecting vast areas over several days, with little suffering and few mortalities. It is rare, indeed, that so many perish in a particular locality excepting the occurrence of a tornado or flood. The twenty-nine-inch snowfall that snuffed out ninety-five lives in Washington on that January evening more than half a century ago is a prodigy that still stands on the record books and was the worst individual tragedy to visit that city.

9.
TWO GREAT SNOWS IN EARLY 1958

Winter came in gently on December 21, 1957, with bright sunshine and temperatures soaring into the fifties throughout a large section of the East and producing average amounts of snowfall in its first two months of life. Nothing exceptional. Come the first of February, cold Canadian air invaded the country, east of the Rockies, and persisted with below-freezing temperatures for the better part of the month. Living in the fifties, knee-high to a fir tree, I came to expect the worst from the weather. After all, the hurricane seasons in the early and middle part of the decade were nothing less than spectacular and the winter of '55-'56 was slow to depart with two early spring snowstorms depositing better than twenty inches along the East Coast in only a matter of three or four days. The late winter of 1958 was not disappointing.

At dawn on February fifteenth, a great blizzard was sweeping eastward, clogging the Great Lake states with the usual winter fare of a good healthy two- or three-day snow, measuring upwards of two feet. As usual, the culprit along the East Coast was not the snow from the lake-effect, but the energy transferred from this cold-weather system to the moisture-laden gulf, breeder of some of the most intense northeasters on the face of the continent. The cold wave that sparked the first major storm of the season in the East was so intense that the Mississippi River sealed itself to vessels from Cairo, Illinois, northward. In fact, it was just a repetition of the weather pattern that was characteristic of the winter as a whole. A strong upper-air low was stalled over the New England coast and a massive high had stationed itself over the center of the country, and the result was that the north wind had easy access to the Midwest and East between these two dominant systems. The difference in February was that the cold intensified and traveled over reflective snow-covered land and didn't have much opportunity to modify on its journey east.

Snow pushed northward from the central gulf states, where the storm had been intensifying as it encountered intensely cold air and the first flakes descended Saturday afternoon in the region from the Delaware Bay to New York City. Over the next twenty four hours, a full-scale blizzard was to develop, as temperatures slumped through the teens and winds increased across the area to forty-five miles an hour on an average with gusts sixty miles an hour and greater. Snow shook out of the sky in an explosion of billions of soft white flakes which quickly blanketed the countryside in a mantle of deep fresh snow. The wind swirled through the wispy accumulation and drove it into huge long banks which were five feet high in the South, ten feet in the Baltimore-D.C. suburbs, and ranged as high as twenty feet in the northeast corner of Pennsylvania where residents of Wayne County were buried under a forty-two-inch cover. The mantle of new snow was not confined solely to the East in this instance, for vast areas stretching from the Mississippi to New England struggled through similar amounts of heavy snow and brutal cold. Michigan City, Indiana, recorded forty inches, a standard that even today is a measure of every snowfall.

Temperatures were on a steep decline throughout the eastern two-thirds of the country as the central and northern plains observed readings of thirty to forty below. As that air streamed eastward, it fed the raging coastal storm that was sweeping the East as readings atop Mt. Mitchell in North Carolina plunged to twenty-seven below. A state of emergency was declared for all of Delaware as snowdrifts blocked every major highway in the First State. Connecticut, with its fifteen-foot drifts, was in a similar condition as the storm dumped a foot and a half of fresh snow. Syracuse, New York, measured 61.1 inches of snow on the ground when the storm passed, a record for February accumulations.

The enormous forces that fueled this storm were evident from the unwinterlike melted equivalents reported at various stations: from Norfolk, Virginia, to Boston, amounts approached three inches or more. Year-round residents along the south Jersey coast were astounded to see the first-ever iceberg emerging from the Brigantine Bay. It measured twenty feet wide by thirty feet long and towered twenty feet above the frozen waters, floating across the bay. This was the first known iceberg in the United States south of the Mason-Dixon line. The storm left 171 dead from Alabama to Maine as a result of exertion and exposure.

The full fury of the winter season of 1958 was far from over. Temperatures resumed their plunge and attained winter lows on the morning of the eighteenth when most stations in the East reported zero readings or below. While temperatures eventually moderated, it was still cold enough to snow in the waning days of winter. On Friday, March 14, a moderate snowstorm dealt a glancing blow to the East Coast, knocking out power and transit systems temporarily, and was just a trace of things to come. The bumpy, frozen remnants of the modest snow still lay on the ground when snow began anew on Wednesday evening, the nineteenth. In many sections, precipitation began as rain, and in the coastal section it remained a mixture of rain, snow, and sleet for the duration of the storm. The equinoctial snowstorm of 1958 is one that almost escaped the record books, for temperatures up and down the East Coast were within a few degrees of freezing. In Baltimore, for example, only seven inches of wet slushy snow was noted, but on the west side of town, up to twenty-eight inches piled the streets with devastating effects on power lines and trees.

Winds increased over a broad stretch, and velocities of forty miles an hour or more were quite common as the coastal storm intensified and began a slow crawl up the coast. Philadelphia proper counted only six to eight inches of new accumulation, but in its northern and western suburbs, the twenty-five inches which fell in West Chester broke a thirty-nine-year snow record. Further west, near the Morgantown Interchange of the Pennsylvania Turnpike, nearly 800 people found themselves stranded in a Howard Johnson's restaurant for nearly thirty-six hours as a result of the surprisingly heavy winter storm. Most had been on the road, and many were forced to abandon their cars and walk several miles through the snow, some carrying small children, to the only refuge for miles around. Panic set in when the group realized that the restaurant was only geared to cater to 100 and food supplies quickly ran low. One saving grace was the appearance of two Amish farmers who distributed their load of baloney and cheese for a modest fee to the adults and freely to the children in the group.

Fortunately, there were three doctors among the stranded, and medical aid was rendered on-the-spot to those who were ill. Army helicopters came to the rescue of the score of people who were seriously ill. Those who remained found it difficult to comprehend why help was so slow in arriving when they were only thirty-five miles out of Philadelphia. One of the big reasons was that nearly forty-two inches of heavy wet snow had accumulated between Downingtown and Morgantown and that twenty-mile section was littered with over a thousand cars and trucks, many of whose drivers decided to ride out the storm in their snowbound vehicles. This tactic was the wisest of choices for so many people miles from any shelter, for to travel on foot in such conditions meant almost certain death.

The weary travelers took to sleeping on the floor in shifts, for there was not enough space even in the washrooms and boiler rooms to accommodate everyone at once. They were in this together, and the panic subsided as the group at large became more organized. It was not until 10 A.M. on the twenty-first, nearly two days after their ordeal began, that rescue—in the form of giant bulldozers—began. Those who had come staggering along, single-file, tripping and falling in the snow, and who had lost their shoes to the gigantic drifts, were about to be saved. The army vehicles zigzagged through the stranded autos and trucks and finally plowed a single lane back to civilization, throwing the snow into huge roadside piles as they plowed through. Ironically, in Pittsburgh, just 250 miles down the road, only flurries were experienced in this period. The nightmare was over.

Meanwhile, the storm was earning accolades from the press as it increased in violence all day Thursday, the twentieth, and piled up two to three feet of snow throughout the inland section of the Mid-Atlantic and New England. It was acclaimed to be the worst storm in Pennsylvania in forty years as over a million power customers across the eastern end of the state lost their power, the hot water, and in some cases their water as water companies experienced widespread power failures. Many were reduced to melting snow for drinking water.

A few days later, another storm from the gulf emerged, but this time brought only rain. Minor flooding ensued but the late winter of 1958 proved once again Yogi Berra's sage observation, "It ain't over till it's over."

10.
THE SNOWIEST DAY IN CHICAGO

THREE YEARS OUT OF FOUR, A PHENOMENON KNOWN AS THE JANUARY THAW PUTS IN AN appearance throughout the nation's midsection and eastern coastal plain. Oftentimes, the reason is due to a well developed storm system tracking northeastward from the southern Rockies, bringing a surge of mild southwestern air northward with it. The thaw may last for two to four days, depending upon the speed with which a reinforcing surge of cooler Canadian air can penetrate southeastward. There is a decided preference for this phenomenon to occur between the twentieth and twenty-fourth of the month, and January of 1967 fit very nicely into this mold.

I recall taking a train out of Milwaukee on the twenty-fourth of the month, on a fairly mild, drizzly day. It was unusual for temperatures there to remain above freezing for so long in the middle of winter. When we disembarked in Chicago, some ninety miles to the south, a quickening southerly wind raised nighttime temperatures well into the sixties, and it was clear that some major weather changes were afoot. I left town by plane that evening, and one of the last things I recall was that a violent thunderstorm had taken place, in a peculiar mid-winter pique of Nature. A strong cold front was pressing down from the Canadian border, and temperatures dropped overnight from the sixties into the thirties and remained there the following day. I was glad to be heading south.

Meanwhile, temperatures in the East were headed for the sixties and even seventies, as Baltimore registered a high of seventy degrees, shattering the old record for the day (established in 1952) by an extraordinary seven degrees. With such a temperature discrepancy shaping up, a new and powerful storm system would be encouraged to develop, and so one did over the lee side of the mountains in New Mexico on the twenty-fifth. As it moved northeastward on a track up the Ohio Valley, heavy snow broke out to its north, and rain to its east and south. In fact, the weather turned maliciously violent in the southern states of Louisiana and Mississippi, where tornadoes touched down in Boyce and Colfax (Louisiana) somewhat ahead of their early-spring schedules. Ahead of the storm system, temperatures in St. Louis dropped from a high of seventy-four degrees on the twenty-fourth, to a made-for-snow temperature of thirty degrees two days later.

David Ludlam, in his outstanding *American Weather Book*, states: "Never in the history of the Midwest has such a mass of snow fallen on so many people in such a short time as descended from the skies over Chicagoland on January 26-27, 1967." A search of previous snowfall records for Chicago reveals that this was a once-in-a-hundred-years event, for never had the snowfall in a single storm exceeded twenty inches in Chicago. Snowfall records for both depth and intensity were about to fall throughout central and northern Illinois, Michigan, Kansas, and Missouri. When Chicagoans awoke on the morning of the twenty-sixth, thick wet snowflakes were descending and beginning to accumulate by morning rush hour, driven by a wind that reached a peak gust of fifty-three miles an hour from the northeast. Throughout the day snow fell heavily, and by evening rush hour it exceeded one foot on the ground and was still falling in massive amounts. Ironically, this storm had little effect on places north of Milwaukee, as that city measured only one inch with practically nothing in its northern suburbs, so the snow intensity tapered off rapidly in the course of only ninety miles.

On the morning of the twenty-seventh, the snow tapered off but not before providing an astounding twenty-three inches of snow in twenty-seven hours, and twenty-six inches to Chicago's suburbs. Gary, Indiana, reported a similar amount in their record twenty-four-inch fall, with drifting snow piling up fifteen feet in some places. Kalamazoo reported a record seventeen inches in as many hours and a storm total of twenty-eight inches as northeast winds off Lake Michigan fed the moisture content of the storm considerably. Lansing checked in at an even twenty inches and amounts tapered off as the storm headed eastward, spending most of its energy in the nation's midsection. Directly in line with this snowstorm was Portland, Maine, but residents there picked up only a scant thirteen inches. The storm produced a nightmare for homebound commuters—a routine bus ride of thirty blocks which ordinarily took fifteen minutes—saw more than four hours elapse. One man in Chicago's business district decided to head home "early," leaving Chicago about 2:30 P.M. His usual fifty-five-minute drive to suburban Park Forest took a total of eight hours, ten minutes that night. He was one of the lucky ones, for thousands of stranded vehicles snarled the major arteries, and officials estimated that cleanup took ten times longer than necessary, due to the enormous task of removing so many cars from the road. Seventy-six lives were claimed as a result of this storm, most of them in the Chicago area, and removal costs approached three million dollars. Business in the area claimed a loss of 150 million, largely as a result of goods which were not sold.

In the East, tornado watches went up in northeastern Maryland, the state of Delaware, southern New Jersey, and southeastern Pennsylvania, south of the storm track where the greatest amount of converging currents was taking place. A tornado touched down in Felton, Delaware, and a sheath of ice piled up to the storm's north. In the South, tragedy struck at Cape Kennedy. To a country that was growing accustomed to space successes, the news from Florida on January 27 was indeed grim. Virgil Grissom, who was an early pioneer in America's space program, Edward White (who was the first to take a "spacewalk"), and Roger Chaffee were in the Apollo I going through a day-long simulation of their anticipated February twenty-first launch when the final flakes of snow were descending over Chicago.

They were within ten minutes of concluding the exercise when, at 6:31 P.M., one of the crew members smelled smoke. "Fire...I smell fire," came the chilling words from the space capsule, 215 feet above the launching pad. There was a two-second pause, and then Colonel White shrilled, "Fire in the cockpit." After another three-second pause came the hysterical message, "There's a bad fire in the spacecraft!" In an atmosphere of pure oxygen, fire was something they didn't need. There was another seven-second gap where no voices were heard, only the sounds of three valiant men scrambling, clawing, and pounding on the unyielding latch of their spacecraft in a desperate effort to save their lives. Their final words were, "We're on fire—get us out of here!" After that, television monitors picked up the wall of fire. "There was a flash and that was it," commented a NASA official. The capsule disintegrated under the estimated heat of 2500°F, and a nation mourned the loss of its first astronauts. Ironically, these men died on the ground, despite an impressive safety record of 1900 man-hours spent in space, as thirteen million miles had been logged previously, an equivalent of twenty-five round trips to the moon.

President Johnson declared after the tragedy that this nation "still intends to land the first man on the moon by 1970," a promise that was fulfilled in July of 1969, when Walter Cronkite uttered the words, "Oh boy." In closely related news of the day, the United States, Soviet Union, Britain, and fifty-nine other nations signed a peace treaty vowing the limitation of military activities in outer space. Strangely, the mysterious link between extreme weather events and aircraft disasters were fulfilled on this day, January 27, 1967, when Chicagoans began digging out of their deepest snow and Americans were rethinking their volatile space program. The snows of 1967 eventually dissolved and man did conquer the immediate limits of outer space, but deeper snows will one day fall as the limits of the weather books and man's knowledge of space continue to expand.

11. SOME NOTABLY WHITE CHRISTMASES

IN THE CHARMING STORY OF A CHILD'S CHRISTMAS IN WALES, THE POET DYLAN THOMAS cannot recollect "whether it snowed for six days and six nights when I was twelve, or whether it snowed for twelve days and twelve nights when I was six." Funny how things are that way. Within the past generation, three major winter storms have unleashed their fury in various parts of the nation on Christmas Day, leaving a pristine mantle of snow in just enough time to try out a new sled. Actually, the week between Christmas and New Year's historically brings out the worst in the weather, at the darkest time of the year.

Perhaps the fiercest of all Christmas storms in the period was the howling northeast gale that swept up the East Coast in 1872, which might be described as the whitest Christmas of them all. If there were Flexible Flyers available back in those days, our great-grandparents had to wait a couple of weeks for the sheer quantity of snow to subside. A few days before Christmas, cold had gripped the nation's midsection and the South. Milwaukee notched an even thirty below while Chicago registered a relatively balmy minus twenty. The zero line extended all the way south to Memphis, Tennessee, and many trains came to a standstill, their locomotives frozen in the biting chill. There were innumerable train wrecks, notably in Indianapolis, St. Louis, and near Erie, Pennsylvania, which claimed thirty lives, all weather-related.

Snow began falling in the southern tier, from the Carolinas northward into the Virginias on the morning of Christmas Day. In Columbia, South Carolina, the storm commenced as sleet and turned to snow, bringing that region of the country their first white Christmas in the ninety-five years in which records were available. By dawn on the twenty-sixth, the snow line had penetrated New England, and it had become quite deep throughout the Mid-Atlantic states. As the day progressed, it became "frightfully deep" measuring two feet on a level in Baltimore and New York. Backyards became a broad and unbroken expanse of glistening snow, and a high, piercing wind drove the snow through every cranny and crevice. Snowplows, in those days, consisted of an archaic wooden structure drawn by a ten-horse team. Travel was virtually abandoned as railroad cars struggled all day and floundered in the snow until all the horses were worn out, compelling thousands to walk home after a day's work. It was amusing, however, to see people struggle and stumble along, "but little sympathy was manifested for the downtrodden; on the contrary, loud laughter invariably greeted the luckless individual who measured his length in the snow."

One of the most remarkable aspects of this storm was the darkness of the atmosphere, such that "people could scarcely discern objects a few feet distant." Snow covered the countryside with a poetical whiteness, and had all the charm that a child could wish for in his fondest of dreams.

In 1909 snow got underway promptly at dawn just south of Washington, D.C., and spread heavy wet snow into the Delaware Valley in time for the traditional opening of the stockings. This time, winter bestowed most of its gifts on Philadelphia, setting a record of twenty-one inches which remained virtually unchallenged until February of 1983. The storm was accompanied by winds of nearly sixty miles an hour, which blew down trees, shattered many a window and produced huge drifts, in some places towering over the second-storey windows. One gauge of a successful snowstorm is a comparison of it to the remarkable Blizzard of 1888, and in this case, the City of Brotherly Love had enough snow to make the storm of 1888 look like a mere flurry.

Meanwhile, along the New Jersey coast, ships were being battered about like tugboats in a bathtub. One steamer, the *Thurman*, lay helplessly adrift and sinking off Tom's River, about twenty-five miles north of Atlantic City. In the days before the Coast Guard, brave men of the Signal Corps patrolled the beaches wandering halfway between stations that were positioned about eight miles apart, until they met another corpsman in all kinds of weather and in the cruelest conditions. There they exchanged messages and made the lonely trek back to their stations, both eyes on the sea in search of ships in distress. Christmas Day 1909, the flares from the hapless *Thurman* caught the eye of one, giving the signal to the life-saving crew. There was a near hurricane wind blowing, and the snow and sleet were blinding.

One account relates:

> It seemed like madness for the life-savers to attempt to launch their big boat, but they went at it as though the sun had been shining and the breath of Summer blowing. Time after time, they essayed to get their boat beyond the breakers, but each time they were beaten back by the wind and waves. Finally, they gave up their attempt and they brought out their gun and line and breecher buoy. Three or four times the slender line and its weight were fired and at last it fell over the rigging of the stranded vessel.

In all, twenty-seven half-perished men were brought ashore and taken to the rescue station to thaw out.

The snow finally ended by noon on the twenty-sixth, but not before knocking out power and telegraph lines once more. In Boston, a great tidal wave broke across the Massachusetts Bay, measuring nearly fourteen feet in height, flooding much of Boston. It was truly a tidal wave, as a full moon shone above the storm clouds. This wave was exceeded only by the one produced in the great storm of April 1851, when the Minot Lighthouse was destroyed.

1945 was the first Christmas to be celebrated in peace since 1938, and thousands of American soldiers were greeted on this country's eastern shores by a raging rainstorm that was in progress. This was not a coastal disturbance, as Harry Truman discovered when he flew to his mother's home in Independence, Missouri. The trip took nearly six hours by plane from the Capital, and severe icing and sleet plagued the craft from the onset. It was perhaps the most perilous trip ever undertaken by an American president since Teddy Roosevelt rode out of the White Mountains of New Hampshire on horseback to be sworn in following the assassination of William McKinley.

Two years later, in 1947, the area

from Washington to New England was blitzed in a massive snowstorm beginning in the late hours of Christmas night. New York City fell victim to this storm in the extreme. Before it stopped snowing the following night, Gotham recorded its heaviest fall, a staggering 25.8 inches that warranted a triple-banner headline in the New York *Times*. Unlike the blizzard on Christmas Day of 1872, winds were relatively calm in this instance, and great blobs of snow descended over New York at the rate of one to three inches an hour, bettering the thirty-hour record of 1888, 20.1 inches, in its first twelve hours. New York was enveloped in a cloudburst of snow, a veritable tornado of snow as the surrounding area received greatly reduced amounts from this storm: the Pocono Mountains in Pennsylvania received only twelve inches, nearby Bear Mountain collected eighteen inches, and New Haven a mere fourteen, slightly better than half the total avalanche in New York. Ironically, while the snow slowed down the pace of New Yorkers, to the point that a streetside conversation might be audible, it didn't stop 15,000 people from attending a tennis match between Bobby Riggs and the debut of Jack Kramer in Madison Square Garden that the former won. What it did accomplish in Gotham was an old-fashioned kind of Christmas where oil lamps were restored in the windows and Christmas lights burned through the snow curtain with a multi-colored softness, reflected in the snow.

Finally, in 1966, the famous Donner and Blitzen snowstorm erupted on Christmas Eve, before taking leave by mid-day on Christmas. Interestingly enough, a "cloud bombing" in New Hampshire had failed to precipitate any snow on the bare ski slopes on the White and Green mountains only two days before. Ski resort operators were wringing their hands, as the season should have been getting into full swing over the holidays. Literally tons of dry ice had been dumped into the clouds in hopes that it would cool the water droplets along their path and cause them to crystallize into ice. For this operation to be successful, the crystals must grow and become large enough to fall to the ground as snow.

Meanwhile, some of the first satellite photographs of the weather were being released; upon close inspection, a growing coastal storm moving up the East Coast was revealed. It whitened the ground all the way from the Tennessee Valley and the southern Appalachians to Canada, accompanied by unusual claps of thunder and flashes of lightning. Storm totals throughout the region averaged one foot except in Upstate New York where nearly twenty-one inches accumulated. It was the first white Christmas in the East in five years, but it was well worth waiting for.

Winter was far from over in the East, and many locations set records the following April twenty-seventh, when a touch of snow to several inches fell, marking the latest snowfall in modern times.

Many dream of a white Christmas, and some of these dreams are fulfilled. But occasionally, as history shows, one's dreams turn into a monstrous waking nightmare. It's peculiar that so many noteworthy snowstorms have occurred at the Yuletide that our fancies should yearn for more. Yet each year, we think that Christmas will be that much more exciting and enjoyable if Nature would be so gracious as to provide a white background. The disappointing years aside, it's clear that she doesn't need all that much encouragement.

12. THE BLIZZARD OF 1983

THE LONG AWAITED WINTER OF 1982-83 FINALLY LIVED UP TO EXPECTATIONS. IT HAD BEEN a relatively drab time, especially in light of the widespread notion that this winter would be one for the record books. As early as September of 1982, would-be weather prophets (as well as some genuine meteorologists) had been speculating that the coming season would provide more than enough wintry weather, perhaps the coldest and snowiest of the twentieth century. Even the *Old Farmer's Almanac* intimated that the winter ahead would be nothing less than brutal.

On March twenty-eighth of 1982, El Chicon, an enormous volcano in Mexico, blew its top and spewed millions of tons of sulfuric ash high into the atmosphere. The monster cloud circled the globe in less than two weeks, closing down the world's largest observatory—Kitt Peak—in Arizona and dimming the incoming sunlight in Hawaii by some 33 percent. Atmospheric scientists were quick to note that it was precisely this sort of phenomenon that caused the Backward Summer of 1816, when frosts were noted in every month of the year. To compound matters, the sun was entering a low period in the sunspot cycle, the planets were aligned to produce the greatest gravitational pull upon the earth (which is linked to increased storminess), and a study of temperature and precipitation cycles suggested that the lowest temperatures and the greatest periods of storminess would coincide with the winter season.

Thus the stage was set. Temperatures in November were unusually mild, with the heat wave lasting into the first week of December. Up and down the East Coast, and extending well into the continent, records were shattered as temperatures soared well into the sixties and even seventies. It was not until the second week of December that temperatures settled back to their seasonal norms, but not before New York City recorded an unseasonable high of seventy-one degrees on the third and fourth of the month.

A snowstorm from the South extended into the Mid-Atlantic states and dropped six to ten inches before diminishing. The snow depth line fell off sharply, as Philadelphia International Airport reported 7 inches while nearby Allentown received nothing at all. For many, this was the only wintry event in a lackluster season, until the eleventh of February. Temperatures began to climb again, topping out in the sixties for Christmas Day, and remained mild until New Year's. For many easterners, December comprised one of the warmest twelfth months in history, and many individual high temperatures were set.

Ice and snow buildup at the Blackhawk Spring, Middletown, Pennsylvania. (*Ron DerMarderosian*)

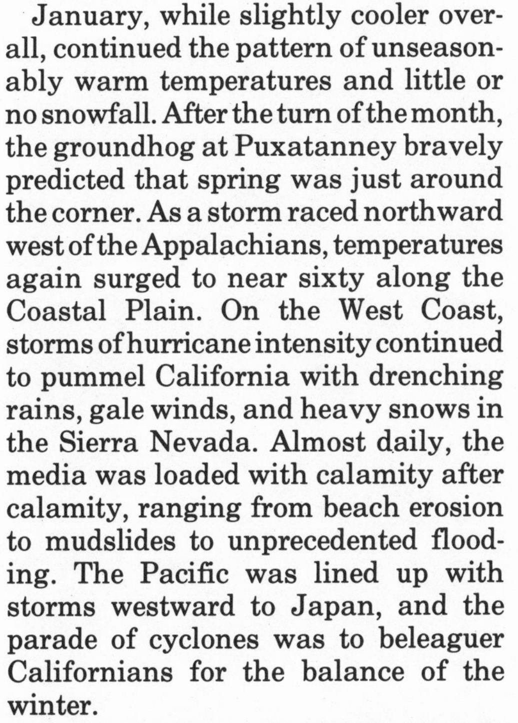

January, while slightly cooler overall, continued the pattern of unseasonably warm temperatures and little or no snowfall. After the turn of the month, the groundhog at Puxatanney bravely predicted that spring was just around the corner. As a storm raced northward west of the Appalachians, temperatures again surged to near sixty along the Coastal Plain. On the West Coast, storms of hurricane intensity continued to pummel California with drenching rains, gale winds, and heavy snows in the Sierra Nevada. Almost daily, the media was loaded with calamity after calamity, ranging from beach erosion to mudslides to unprecedented flooding. The Pacific was lined up with storms westward to Japan, and the parade of cyclones was to beleaguer Californians for the balance of the winter.

One such storm reformed in the Gulf of Mexico and churned northeastward on the sixth of February with snow, sleet, and freezing rain in coastal areas while dumping nearly a foot of snow on inland locations. It was a taste of things to come, for on the morning of the tenth, a very powerful snowstorm was getting its act together in the same region, and this time a cold dome of arctic air was anchored over northeastern Canada and served to intensify and prolong the inevitable precipitation.

As the storm churned slowly up the Atlantic, heavy wind-driven snow buffeted the East Coast, falling at the rate of nearly two inches an hour on the eleventh in the Mid-Atlantic states, and well into the afternoon hours of the twelfth in the northeastern tier. In what was labeled a blizzard, temperatures slumped into the teens and single digits while winds of near hurricane force buffeted the coast and gales swept inland, accompanied by blinding snow.

In the afternoon hours of the eleventh, snow piled up so rapidly as to bring transit and commerce to a complete standstill before winds diminished in the evening.

On that memorable Friday night, it became apparent that nearly every snowfall record would fall, and by midnight they did. Philadelphia, Baltimore, and Harrisburg recorded their heaviest snowfalls in 135 years of record keeping, while every major city from Richmond to Boston measured snow in the one-to-two-foot range. So the storm swirled out to sea over the Canadian maritimes, and left in its wake a blanket of snow so deep as to be a once-in-a-lifetime phenomenon.

At the height of the blizzard, there was a vivid flash of lightning and a clap of thunder that would send one scurrying, even in mid-summer. I was playing Scrabble with my wife and was about to lay down a seven-letter word at the time. Moments later, a friend who I suppose thought he might owe me one, called and suggested that I "make it two, if you think of it." Very much like the Stroh's beer commercial which shows a determined, thirsty man preparing to walk two hundred miles through a raging blizzard for a beer, many easterners—with and without refreshments—had a taste of a full-fledged blizzard in the otherwise tepid winter season of 1983.

13.
WORDS OF WARNING AND WINTER STORM SAFETY RULES

The National Weather Service issues watches and warnings whenever hazardous winter weather is threatening.

A winter storm watch means that severe winter weather conditions may affect your area.

A winter storm warning means that severe winter weather conditions are imminent.

An ice storm warning is issued when significant, possibly damaging, ice accumulation is expected. Freezing rain (or drizzle) means that precipitation is expected to freeze upon contact with exposed surfaces.

A heavy snow warning implies a snowfall of at least four inches in twelve hours or six inches in twenty-four hours is expected. Heavy snow can mean lesser amounts where winter storms are infrequent.

A blizzard warning means that considerable falling and/or blowing snow, with winds of at least thirty-five miles an hour, are expected for several hours.

A severe blizzard warning means that considerable falling and/or blowing snow, winds of at least forty-five miles an hour, and temperatures of ten degrees or lower are expected for at least several hours.

A high wind warning means winds of at least forty miles an hour are expected to last for at least one hour.

WINTER STORM SAFETY RULES

Keep ahead of the winter storm by listening to the latest weather warnings and bulletins on radio and television. A specially equipped weather radio is available and provides continuous up-to-date broadcasts.

- Check battery powered equipment before the storm arrives. A portable radio or television set may be your only contact with the world outside the winter storm. Also check emergency cooking facilities and flashlights.
- Check your supply of heating fuel. Fuel carriers may not be able to move if a winter storm buries your area in snow.
- Check your food and stock an extra supply. Your supplies should include food that requires no cooking or refrigeration in case of power failure.
- Prevent fire hazards due to overheated coal and oil-burning stoves, fireplaces, heaters or furnaces.

  Stay indoors during storms and cold snaps unless in peak physical condition. If you must go out, avoid overexertion.
- Don't kill yourself shoveling snow. It

is extremely hard work for anyone in less than prime physical condition, and can bring on a heart attack, a major cause of death during and after winter storms.

- Rural residents: Make necessary trips for supplies before the storm develops or not at all; arrange for emergency heat supply in case of power failure; be sure camp stoves and lanterns are filled.

Dress to fit the season. If you spend much time outdoors, wear loose-fitting, lightweight, warm clothing in several layers; layers can be removed to prevent perspiring and subsequent chill. Outer garments should be tightly woven, water repellent, and hooded. The hood should protect much of your face and cover your mouth to ensure warm breathing and protect your lungs from the extremely cold air. Remember that entrapped, insulating air, warmed by body heat, is the best protection against cold. Layers of protective clothing are more effective and efficient than single layers of thick clothing; and mittens, snug at the wrists, are better protection than fingered gloves.

Your automobile can be your best friend or worst enemy during winter storms, depending upon your preparations. Get your car winterized before the storm season begins. Be equipped for the worst. Carry a winter-storm car kit, especially if cross-country travel is anticipated or if you live in the northern states.

Suggested Winter-Storm Car Kit: blankets or sleeping bags, matches and candles, empty three-pound coffee can with plastic cover, facial tissue, paper towels, extra clothing, high-calorie nonperishable food, compass and road maps, knife, first-aid kit, shovel, sack of sand, flashlight or signal light, windshield scraper, booster cables, two tow chains, fire extinguisher, catalytic heater, axe.

Winter travel by automobile is serious business. Take your travel seriously.

- If a storm exceeds or even tests your limitations, seek available refuge immediately.
- Plan your travel and select primary and alternate routes.
- Check latest weather information on your radio.
- Try not to travel alone; two or three persons are preferable.
- Travel in convoy with another vehicle, if possible.
- Always fill gasoline tank before entering open country, even for a short distance.
- Drive carefully, defensively.

IF A BLIZZARD TRAPS YOU

- ***Avoid overexertion and exposure***. Exertion from attempting to push your car, shovel heavy drifts, and perform other difficult chores during the strong winds, blinding snow, and bitter cold of a blizzard may cause a heart attack—even for persons in apparently good physical condition.
- ***Stay in your vehicle***. Do not attempt to walk out of a blizzard. Disorientation comes quickly in blowing and drifting snow. Being lost in open country during a blizzard means almost certain death. You are more likely to be found, and more likely to be sheltered, in your car.
- ***Don't panic***.
- ***Keep fresh air in your car***. Freezing wet snow and wind-driven snow can completely seal the passenger compartment.
- ***Beware the gentle killers***: carbon-monoxide and oxygen starvation. Run the motor and heater sparingly, and only with the downwind window open for ventilation.
- ***Exercise*** by clapping hands and moving arms and legs vigorously from time to time, and do not stay in one position for long.
- Turn on dome light at night, to make

the vehicle visible for work crews.

- ***Keep watch***. Do not permit all occupants of the car to sleep at once.

Courtesy of U.S. Department of Commerce, National Oceanic and Atmospheric Administration, National Weather Service

---

The word *snow* in a National Weather Service forecast, without a qualifying word such as occasional or intermittent, means that the fall of snow is of a steady nature and will probably continue for several hours without letup.

Snow flurries are defined as snow falling for short durations at intermittent periods; however, snowfall during the flurries may reduce visibility to an eighth of a mile or less. Accumulations from snow flurries are generally small.

Snow squalls are brief, intense falls of snow and are comparable to summer rain showers. They are accompanied by gusty surface winds.

Blowing and drifting snow generally occur together and result from strong winds and falling snow or loose snow on the ground. *Blowing snow* is defined as snow lifted from the surface by the wind and blown about to the degree that horizontal visibility is greatly restricted. *Drifting snow* is used in forecasts to indicate that strong winds will blow falling snow or loose snow on the ground into significant drifts. In the Northern Plains, the combination of blowing and drifting snow, after a substantial snowfall has ended, is often referred to as a ground blizzard.

Blizzards are the most dramatic and perilous of all winter storms, characterized by strong winds bearing large amounts of snow. Most of the snow accompanying a blizzard is in the form of fine, powdery particles of snow which are whipped in such great quantities that at times visibility is only a few yards.

III. TORNADOES

In the original plans for this book, there was to be a single chapter dealing with violent storms; i.e., tornadoes and hurricanes. Since then, I have gained a world of respect for the relatively small local storm which has swarmed through many a Midwest town on hot sultry days. The tornado is unique in so many ways and while it favors the Midwest, there are very few places east of the Continental Divide that have not witnessed their consuming fury at one time or another. While the tornado has its season—April, May, and June—there is only one day of the year (January sixteenth) in which no tornado has ever been observed in the forty-eight conterminous United States since 1950. On the other hand, every state in the Union including Alaska and Hawaii has experienced at least one tornado in the past thirty years.

Tornadoes are virtually impossible to predict two or three days ahead of time with the confidence of a meteorologist foretelling a general rain or snowstorm. Weathermen are familiar with the conditions which will favor tornadic development and hence we frequently hear of watches *versus* warnings. *The distinction is made by the National Severe Storms Forecast Center (NSSFC) in Kansas City, Missouri. This branch of the National Weather Service is responsible for defining a watch area, which is approximately 100 miles wide by 250 miles long where weather conditions suggest a high probability of tornado activity. These watches also specify a time frame where tornadoes have the highest probability of occurring. Watches are teletyped directly to the local National Weather Service offices and disseminated to the general public via commercial radio, television, and NOAA Weather Radio. A watch also activates law-enforcement officers, civil-defense personnel (such as the Red Cross) and specially organized radio-spotter groups.*

A watch is not a warning. A watch *means that tornadoes are possible, both within and near the watch area, and people should be on the lookout for threatening conditions and stay advised of further developments. Otherwise, activities should not be disrupted. A* warning, *on the other hand, means a tornado has been spotted and is in progress for the affected area, and immediate action should be taken (see the appendix of this chapter for spe-*

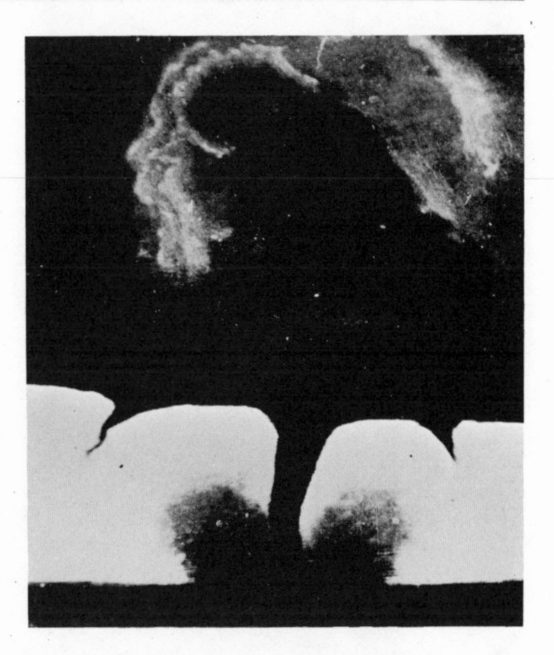

The first known photograph of a tornado, taken twenty-two miles southwest of Howard, South Dakota, on August 28, 1884. (*National Oceanic and Atmospheric Association*)

cifics). The area of the warning is determined by the location, size, direction (which can be erratic), and speed of movement of the storm. Since tornadoes are not always indicated by radar or seen, a warning may not always be given. When warnings are received, persons close to the affected area should take cover immediately.

On the average, tornado tracks are a quarter-mile wide and seldom touch down for more than fifteen miles. In extreme cases, these storms have been over a mile wide and trekked several hundred miles across four or five states. Forward speeds are usually about thirty miles an hour, but some have been known to travel seventy miles an hour or better, reducing the time to take shelter. It is this type of storm which is responsible for most tornado-related deaths. More than ninety percent of these storms move from a western quadrant to the east, and most of these approach from the southwest. In other words, if you are to the south of an observed tornado, chances are very good that you are safe. Another rule of thumb is that if you observe the entire tornado from a distance, it will not be passing directly overhead. However, if it is only partially visible, and clouds seem to be lowering, it may be coming directly toward you.

The mathematical possibility that a specific location will experience a tornado in any given year is rather slim. Even in the most tornado-prone country—the Midwest and South—the odds that a storm will strike a given point in the area are only on the order of once every 250 years or so, but in the far western states the probability is near zero. However, the tornado is an exceptional phenomenon and Nature has provided quite a history of unmathematical events. Thanks to a compilation by the National Weather Service, we now know that Oklahoma City

A dramatic waterspout, churning the waters off Martha's Vineyard, five and one-half miles at sea, is seen in this photograph shot through a droplet-filled window. This waterspout reached a height of 3,600 feet and formed on three separate occasions on August 19, 1896. (*National Archives*)

has been struck by tornadoes no fewer than thirty-two times since 1892. The town of Baldwin, Mississippi, was struck twice by tornadoes during a twenty-five-minute interval on March 16, 1942, and Austin, Texas, also endured a similar fate on May 4, 1922. For those who believe that bad luck happens in threes, Codell, Kansas, was struck in 1916, 1917, and 1918—each time on May twentieth.

While Dorothy's adventure in The Wizard of Oz *may have been fantasized, tornadoes have trifled with mankind in less subtle ways. For example, letters have been carried more than a hundred miles and carried people and animals through the air for several hundred yards, or more. In 1931 a Minnesota tornado carried an eighty-three-ton railroad coach and its 117 passengers eighty feet before throwing it down, and in 1975 a Mississippi tornado carried a commercial freezer more than a mile. There are even instances when tornadoes have struck when the sky was virtually cloudless.*

Finally, a note about preparedness. On May 6, 1975, what was to become history's most destructive tornado struck the city of Omaha, Nebraska, at about 4:30 P.M. *on a weekday. It was estimated that 31,000 people lived or worked in the vicinity which caught the full fury of the storm (a downtown section of about 200 city blocks), and winds were measured in excess of 200 miles an hour. Thanks to a network of early warnings—namely from the Weather Bureau working hand in hand with CB radios and civil defense officials—sirens blasted about ten minutes before the storm began its sixteen-mile path of destruction, and only three lives were lost. More than 2,000 homes and businesses lay in total ruin, but when the alarm sounded, people knew what to do and did it.*

14.
AN INCONCEIVABLY TERRIBLE STORM

TOWARDS THE END OF THE FIRST WEEK IN AUGUST, 1878, THE WEATHER FROM THE Mid-Atlantic states north to New England was typically warm and muggy. Temperatures ranged in the eighties and nineties and a humid air mass spawned afternoon thunderstorms, which is not unusual for the region in this season. Providence, Rhode Island, picked up more than four inches of rain in a one-hour deluge, as violent thunderstorms and high winds raked the area. In some communities, it had rained quite hard causing local flash flooding while in others, there was scarcely a drop of precipitation. The community of Wallingford, Connecticut, which lies nestled along the Quinnipiac River twenty miles north of New Haven, had been spared any violent weather and was enjoying one of the loveliest days of the season on the morning of August ninth. Light cumulus clouds dotted the horizon and sailed across the sky until 3 or 4 P.M. It was then that a few black clouds, portending a storm, appeared over Mt. Lamentation, a large hill to the west of the city. Within a few hours, the city would be reduced to rubble.

Lake Windermere lay at the foot of the mountain, a huge artificial pond created by damming the Quinnipiac River. By 5 P.M. workers were leaving their jobs for the week, and many were met by wives and children for the walk home. It was teatime, and children played unaware in the streets. By 5:30, the clouds had become intensely black and people scurried indoors and the little ones pressed their faces against the windows to watch the approaching storm, but there was still no sense of impending danger. The windmill at the base of the pond began to stir, in the quickening breeze. Finally, there was a dead calm just before 6 P.M. but, aloft, the clouds went tearing wildly along at such a pace that most people took to shelter.

With horrible suddenness, a change took place. Under the threatening clouds that had spread over the village like a black curtain, other and blacker clouds seemed to be driven down. The air exploded in a whistling wind, and a low whine was heard from the mountain. A dense cloud mass with a fleecy, misty curtain advanced toward the village, hiding the orchards and vineyards as it moved. Lightning flashed in blinding forks from this cloud, and the flashes lit up the surrounding country with lurid purple light while the thunder rattled and boomed with interminable and deafening loudness.

At 6:15 P.M., a second mass of clouds approached on a northerly wind current, and a countercurrent set in from the south, bringing in heavy black clouds that seemed to touch the treetops. Before a cry could be raised, a death-dealing blast fell upon the people of this poor town.

Observers say that the waters of Lake Windermere were churned into a foam and a great volume of water was lifted from the lake in a swirling motion to a height of 250 feet. With the speed of lightning, this waterspout headed east, toward Wallingford. On the flat plains between the lake and the town, there lived several black families who were poor laborers and built their castles on sand. One account reports that an entire family, living on the plains, were sucked from their house and hurled above telegraph wires and carried through the branches of trees before they were thrown to the ground, battered and gashed, 150 yards away. Entire houses in town were crushed, and fragments were carried aloft in the consuming gale, to mingle with tree trunks and branches, clothing and chairs, and "all manner of things in the craziest confusion." All this happened in less time than it takes to read the grim details.

In twenty minutes, the wind stopped, the rain abated, the sky cleared, and night came on. The village of Wallingford lay in ruin, and the Town Hall bell was rung. The old school house was converted into a morgue where thirty-two bodies lay in simple pine boxes, packed with ice, and the water that dripped from them collected in a pool nearly one inch deep. One man, whose family was wiped out and was repeatedly drunk, uttered:

> The family is all gone...but the dog ain't dead yet, the dog's time didn't come yit." He shrugged his shoulders as if a thought of the dreadful truth had penetrated his understanding. "It's pretty hard and you can't tell why the dog was left and the others were kilt, but I know a few things. The reason why the dog wasn't tuk was the dog's turn didn't come...the dog's turn 'ill come soon enough.

One of the most peculiar aspects of this storm was that, in a mixed community, every person killed was a member of the town's Catholic church.

15. THE GREAT ST. LOUIS DISASTER OF 1896

TORNADOES ARE NOT EXACTLY UNKNOWN TO THE MIDWEST. IN FACT, OF THE THOUSAND OR so which strike the United States every year, a majority of them swipe the Midwest and South. What is unusual, even in the spring of 1896 when St. Louis was a growing city of half a million, is for a major tornado to cross a large metropolitan area.

For two weeks in May of that year, tornado outbreaks had been widespread throughout the region, devastating areas from Iowa, Kansas, Illinois, and as far east as Virginia. On May twenty-fifth and twenty-sixth, tornadoes leveled the Iowa towns of Polk, Jasper, Valeria, and Ankeng. One tornado struck the city of Santiago with unprecedented fury, lifting a house bodily from its foundation and carried it some distance before throwing it to the ground in a shattered wreck, killing an entire family. On a fearfully hot sultry day, rain and thunderstorms broke out and swept across the country at a terrific speed. While the wind was whistling through the trees and sheets of rain pelting the countryside, a great black funnel hung down from a blacker heaven and tore across the prairie faster than the speediest express train. Then it happened. Suddenly, the awful roar that every prairie farmer knows as the dread forerunner of a cyclone, was heard. In Valeria, a house which stood on a high hill overlooking the surrounding county was enveloped by the pendant. Most people survived in cellars and caves when they heard it coming; the house and its occupants were scattered so far that no trace could be found. The storm pursued a path through Oakland County, in Michigan, leaving forty dead in a sixteen-mile rampage.

In the afternoon of the twenty-seventh, the western horizon was banked with clouds which seemed to pile one on top of another with curling yellow-tinged edges. The sky suddenly darkened and funnels just shot out, some projected into the air, and others leaped to the earth, twisting and turning. Lightning played about them, and there was a marvelous electrical display before the outburst. At breakneck speed, three tornadoes approached St. Louis, one from the northwest, one from the west, and another from the southwest. Their descent was so sudden that people fleeing were caught in the streets and hurled to destruction, instantly, or buried under falling walls. From the clouds came a strange crackling sound that filled the air with a stronger hiss than the incessant peals of thunder. The funnels merged into one and enveloped the west side of the city. In thirty minutes, the destruction was total.

St. Louis at the time was a fledging city, the leader in railroad traffic for

Union Depot Railroad Company's power house, showing the massive smokestack that snapped in two. (*National Archives*)

the country, and in a few days was to hold the Republican Convention that was to nominate William McKinley. St. Louis boasted one of the sturdiest bridges in the world, the Eads, which led to the east side. This bridge also housed the Terminal Railroad through which trains passed on the journey east and south. When the storm struck, two trains—one heading east and the other west—were gripped to the tracks of this solid structure. The Wabash Fast Train was the more fortunate of the two. The engineer of that train reported, "The storm was coming squarely in at us. The air began to fill with loose boards and debris. I stopped the train and crouched behind the boiler. The train ran backwards from the full force of the storm...the rear car was on a curve and caught the full force of the storm and shook like a leaf. What saved the train was that it pointed directly into the gale." Eastbound, the Terminal train was completely overturned and wrecked against the framework of the bridge. The bridge itself was twisted into a pretzel on the east end where the Vandalia Freight House stood. That building was blown into atoms and all but four employees were killed.

In St. Louis proper, 720 city blocks were totally demolished, and every rooftop in the storm's path was torn off as if held on with thread. Telegraph poles toppled in long rows, not one by one but a dozen or more at a time. Lightning struck the Standard Oil works, and flames were soon pouring from a dozen buildings. The eastern sky was ablaze as a great city burned. The rain ceased for a time, but by 7:30 P.M. fresh new torrents extinguished the columns of fire. Along the Mississippi waterfront, nearly every vessel was wrecked, and the water was filled with people struggling for life.

The reign of terror passed east, and the following day tornadoes reformed as far east as Ambler, Pennsylvania, where more than seventy-five barns and houses were completely destroyed. The storms jumped the Delaware and died out in Mercer County, New Jersey, after killing more than 500 and injuring thousands.

16.
WESTERN TORNADO IN THE EAST

EARLY SPRING PROVIDES THE EAST WITH SOME OF THE YEAR'S WILDEST WEATHER EVENTS, and a recent study by the National Weather Service indicates that spring is the period when the greatest number of storms occur, especially in its first full month. One need only to read ahead to the story concerning the deadliest tornadic outbreak of the century or note when some of the worst floods and snowstorms occurred in our meteorological history. Moist warm air from the Gulf of Mexico is asserting its seasonal right to flow north, while there is always a crisp supply of Canadian air waiting to spill over into the northern and middle latitudes. This conflict in air masses results in strong atmospheric turbulence which occasionally leads to tornadoes.

A blizzard was in progress in Washington County, Pennsylvania, which houses the suburban communities of Pittsburgh. Temperatures there had dropped more than thirty degrees over an eight-hour period, and drenching windswept rains quickly turned into a blinding snowstorm. Along the Eastern Seaboard, temperatures had become quite mild for so early in the spring, and little did the millions of people living in a broad swath from Philadelphia to Boston suspect that the day's crowning touch would end in ruins for so many. The contrary forces of winter and summer were at work as the first full-fledged tornadoes of the century were about to descend.

Philadelphia and adjacent New Jersey were hard hit by the events of March 27, 1911. The suburb of Tacony, twelve miles up the Delaware from Philadelphia, was nearly blown into a different longitude. The entire Pennsylvania Railroad station was hurled into nearby fields, while shelter sheds and signal bridges were thrown from the tracks. In nearby Wissinoming, freight cars tumbled like autumn leaves from their tracks and twisted into knots. A wrecking train and crew was dispatched from Philadelphia equipped for repair, and it met a similar fate. In the relative calm of the city center, it was hours before dispatchers learned of the train's demise.

Most roofs in the Kensington section of the city were removed, and tombstones were picked up by the wind and carried through the air several hundred feet. To the north, in Pottsville, hail descended so heavily as to coat the ground six to eight inches. In Coatesville, one stone measured 5¼ inches in circumference, a record for the area. High winds and hail blew down barns and smashed every window from York, where a barn was transported 150 feet where it was hurled against a power transmission line, to Burlington, New

Jersey, where many barns were blown into the Delaware River. Burlington was especially hard hit as nearly every building within a five-mile radius lost its roof and there were many hair-breadth escapes from flying timbers and live wires.

Even New York City experienced a baby tornado, and much of the area from Fifteenth to Twentieth streets, river to river, was destroyed while teetering walls presented quite a hazard for city dwellers. While tornadoes in any metropolitan area—and especially in the East—are extremely rare, the storms which wracked the East Coast in 1911 proved that they can occur anywhere, anytime.

17.
EVERYTHING WAS THERE BUT THE HOUSE — THE DEADLIEST TORNADO

On the morning of Wednesday, March 18, 1925, a series of events occurred in Missouri's Ozark Mountains that was to touch the lives of tens of thousands in a matter of hours. A tornado was born there, destined to destroy thirty-six towns over a three-hundred-mile swath covering five states in a matter of only five hours. The storm was the deadliest of its kind in American annals to date. Nearly 700 people were killed with 3,000 injuries and 15,000 homes destroyed. This towering, mile-wide tornado cut a path of devastation through Missouri, Illinois, Indiana, Kentucky, and Tennessee. One survivor, Charles Boggs, a laborer from Murphysboro, Illinois, whose town was struck with the full fury and was demolished and burned, related the following story:

> I jumped from my car when I saw the tornado coming...the machine turned over a couple of times and I never saw it again. I walked the rest of the way to my home, a mile and a half. The place was a complete wreck. I found my daughter-in-law sitting up dazed and she died while I tried to talk to her. Her 2 daughters were 25 feet away, dead. My wife and mother were there too, dead. Nearby was my 21 year old son, Fred. He was dead. My other daughter, at school, was the only one saved. I was only scratched.

When the storm moved through West Frankford, Illinois, around 3 P.M., the greatest toll was claimed as over 1,000 people were seriously injured or lost their lives. The wind was so great that some of the bodies were blown one and a half miles out of town. There were shouts from a group surrounding a distant pile of debris that had a short time before been a comfortable dwelling. A miner, still in his pit clothes, came toward the crowd with a pitiful little bundle, limp in his arms. The bleeding figure, a five-year-old, had been taken from the wreckage of his father's home more dead than alive. An eyewitness reported that "he made no movement but the miner said he was alive." Elsewhere, people lay beneath the recesses of fumbled timbers, and cries came from the injured persons who were pinned beneath the wreckage, while bodies of the dead could be seen far down in the debris, where it was impossible to pull them out. Forty of the dead were babies.

In De Soto, Illinois, the tornado struck a two-story brick school house where there were 125 students and teachers in the building at the time. Only three escaped death. There was no building in town left standing more than ten feet high.

Other school children were more fortunate. In Cape Girardean, Missouri, near the tornado's birthplace, a school house was carried forty feet only thirty

A remarkable photo of a tornado in progress in Minnesota (1925). (*LIL*)

Tornado damage in Murphysboro, Illinois on March 18, 1925. Note that the home in the center background is unscathed. (*National Archives*)

The track from the tornado's damage extended from the Mississippi River to beyond Princeton, Indiana—133 miles. The death toll was 689, and 1,980 were injured. Here is the devastation at DeSoto, Illinois. (*National Archives*)

Residents of Princeton, Indiana, look over the ruins of the C & EE Railroad tracks, which shows the completeness of the storm path. (*National Archives*)

Some of the *least* damaged houses left in the village of Griffin, Indiana, where not one structure was left for occupancy. Approximately 55 people were killed, and 200 were injured out of a population of 350. (*National Archives*)

minutes after the pupils were dismissed for the day. A mother in De Soto was lying in bed with her two-week-old girl when her home collapsed and the timbers fell across the bed in such a manner as to spare her and the baby.

The pictures of devastation slowly came from towns isolated by broken communications and fire. Here, a photograph of an old man and woman staring vacantly into the pit of what was only a couple hours ago their home. Their arms were not around one another. Instead, they stood apart, bent and gray, so hopeless that neither could add to the other's burden by a touch of tears. The air is charged with the cries and screams of those who have lost homes and friends. Homeless, helpless, they stand stricken by the calamity before them. One by one, the lights in the kitchens of those spared flicked on as evening spread its soft wings over the wounded towns.

Massive relief from the American Red Cross poured in by rail from all over the country. Large boxcars carrying as many as 200 doctors and nurses and several cars of medical supplies and tents rolled in from Chicago, St. Louis, and Indianapolis to care for the injured and bury the dead. It was a world-wide event with condolences and relief messages coming from Japan, Canada, the Vatican, and most European countries.

The path of the tornado varied in width from a few yards to a deadly one-mile swath. Fate's hand dragged a path of death across our fair land, hitting here, jumping there in a serpentine course that devoured all in its wake. A crash of thunder, a few flashes of blinding lightning and there was

A jumble of furniture, household goods, and other debris left by the tornado at Griffin, Indiana. Such scenes were the rule rather than the exception at this little village. (*National Archives*)

nothing left but whole blocks reduced to kindling and "the pitiful little black clusters of grieving kinfolk in front of the morgues."

Freakishly, a letter wearing the postmark Poseyville was carried clear to Bloomfield, more than 100 miles away, so tremendous was the force of the wind. One of the few who escaped in West Frankford did so by climbing into an office safe, to flee the tumbling walls and raging fire which ensued. No words could do this tornado justice.

18.
THE REMARKABLE WORCESTER TORNADO OF 1953

TORNADOES ARE A TRUE RARITY IN NEW ENGLAND, STRIKING ON AN AVERAGE OF ONCE A year, but a truly lethal one occurs only every twenty-five or thirty years. The events of June 9, 1953, are unprecedented in New England's weather history.

Actually, the tornadoes were born in the Midwest, where on the previous day six tornadoes smashed through southern Michigan and northwestern Ohio. 139 people lost their lives that day, 111 of them in Flint, Michigan. Whole families were wiped out, pinned beneath a rain of bricks and timber. Flint authorities reported that "we're recovering bodies faster than we can count them." More than a thousand residents were hurt as the cyclone smashed homes into firewood and felled trees over a 300-foot swath.

June ninth was a day of celebration in England but a day of infamy in New England. Two members of a British team successfully conquered Mt. Everest on their third attempt in conditions somewhat better than they were for 1500 residents of central Massachusetts. Oddly enough, the residents of central Massachusetts didn't know what hit them for more than two hours that afternoon. The event went unreported for some time, as a deadly windstorm plunged southeast through the center of New England. Huge cumulonimbus clouds blackened the afternoon sky, spewing dense rainfall and large hailstones across the countryside. To present-day New Englanders, tornadoes are something of which they read in newspaper dispatches from the Midwest or see the devastating aftermath on newsreels. To the contemporary New Englander, the Worcester Tornado of 1953 was simply unbelievable.

The first sign of this tornado was at 4:30 P.M. as a funnel cloud descended on Petersham, where homes were decimated as though from an explosion from inside. The storm intensified, expanding to a width of nearly a mile in some places as it ran its grim race of death through the towns of Rutland, Barre, and finally to Worcester. Almost as soon as it began, the storm ended, leaving behind scenes that its victims could never forget. Overturned automobiles were as helpless as upside-down beetles, and trees were shorn of their leaves as though invaded by an army of gypsy moths. Perhaps hardest hit was Assumption College, in Worcester, which was reduced to rubble as if from the aftereffects of a bombing raid. It's remarkable that there were no more fatalities than there were from this storm, for it came with appalling suddenness, and the damage, in most cases, was unqualified.

Its assault included the main street in Holden and the northern sections of

Worcester. The storm met its demise sometime after smashing through the little community of Fayville at about 5:00 P.M. but not before taking ninety lives, and injuring more than 1300. There was nearly sixty million dollars worth of damage done and many thousands of lives were disrupted or broken.

In retrospect, only the all-New England hurricane of 1938 took a heavier toll in lives and property in all of the region's storm history. The 1953 storm obliterated more than 2500 homes over a sixty-mile sweep, and it's unlikely that anyone alive today who witnessed cars being whisked about like toys before being flattened will ever forget the black cloud that descended on the town of Worcester on that fateful afternoon.

19. THE LARGEST TORNADO OUTBREAK OF A CENTURY

"TORNADO ALLEY" IS AN APPROPRIATE TERM FOR THE AREA SUBJECT TO THE MOST PROLIFIC tornadic outbreak of the present century. Between April third and fourth, 1974, no fewer than 148 tornadoes descended from the skies in a twenty-four-hour period. These twisters affected every state from Michigan south to Mississippi, from Illinois eastward to New York state. Over that stretch, there were 316 storm related fatalities and more than 6,000 serious injuries. 27,590 families were homeless when the sun came out again and damage surpassed $600,000,000. Xenia, Ohio, and Brandenburg, Kentucky, were all but blown off the map in the ensuing violence.

It was the day Patty Hearst appeared on most front pages brandishing a machine gun and announced that she was joining the terrorist S.L.A. group, which two months earlier had kidnapped her. Out to the west, a strong cold front was driving east into a very warm, humid air mass. Strong thunderstorms broke out 200 miles ahead of the front, and a large egg-shaped mass of cold air aloft squeezed the moist gulf air for all it was worth. Meteorologists understand the big picture, but few can explain the particulars of a tornadic outbreak on this scale. One suggested it was like a "dancer pirouetting on a rotating platform mounted on a truck moving at eighty miles an hour." It was with that velocity that a huge tornado descended upon Xenia, Ohio, and precipitated the greatest disaster ever known in that town. The storm struck at 4:40 P.M., as the town clock bore out. Three days later, its hands were still frozen to that fateful hour.

Eight houses, a fire station and a metal-frame warehouse were lifted from the ground and carried about a mile. At least thirty people were killed in Xenia on this most tragic day. It was a disaster too overwhelming to contemplate and a strange fatalism settled over this town of 25,000. It was as though the people got up every morning of their lives and sifted through the splinters of their homes, scraping unrecognizable debris from the streets and sawing up the fallen trees which once provided shade for their lawns. The following day, officials cut off power to the town, as an active breeze trifled with the innumerable dangling power lines.

Next stop was the farming community of Brandenburg, Kentucky, thirty miles west of Louisville. Paula Wright, a twenty-one-year-old housewife, lived in a home near the Ohio River when she saw it coming. Thirty seconds later, five of the seven residents were dead. The tornado left nothing more than shredded lumber and broken bricks, but somehow she and her brother found themselves unhurt but trapped. Mrs. Wright told investigators, "If I live forever, I will never again see anything so horrible as this." One man who was fishing nearby when the storm broke told state troopers, "I looked up and saw houses, automobiles and bodies flying over and into the water." In all, seventy-one people were killed in this community of 1600.

A National Guardsman drove up to the residence of Vernice Simons, who was sifting through the rubble of her belongings and noticed a soft-water tank that had survived. "Do you want that," he asked. "I never want to see that thing again—it almost killed me," was the reply. Mrs. Simons and her husband had crouched in a closet, and the soft water tank landed on top of them, nearly snuffing out their lives. Four other guests in the home perished.

No other storm is more violent than the tornado, nor any so little understood. What we understand is the aftermath: near the edge of the funnel, trees are uprooted like weeds in a prize garden; at the center, buildings explode and railroad cars are tossed about like the wreck of a Lionel.

20. TORNADOES — INDIVIDUAL SAFETY TIPS

WHEN A TORNADO THREATENS, YOUR IMMEDIATE ACTION CAN SAVE YOUR LIFE!

- ***Stay away from windows, doors, and outside walls. Protect your head!***
- In homes and small buildings, go to the basement or an interior part of the lowest level—closets, bathrooms or interior halls. Get under something sturdy. If possible, the southwest portion of your home offers the greatest protection, so you should head there.
- In schools, nursing homes, hospitals, factories, shopping centers, go to the predesignated shelter areas. Interior hallways on the lowest floor are usually best.
- In high-rise buildings, go to interior small rooms or hallways.
- In mobile homes or vehicles, leave and go to a substantial structure. If there is no shelter nearby, lie flat in the nearest ditch, ravine or culvert with your hands shielding your head.

Tornado watch: Tornadoes and severe thunderstorms are possible.

Tornado warning: Tornado detected, take shelter immediately.

Listen to radio, television or NOAA weather radio for latest National Weather Service Advisories.

Tornadoes are only one of a thunderstorm's killer elements

- Lightning is a major killer. Stay indoors and away from electrical appliances (including the telephone) when storms are nearby. Lightning may always strike outside power lines. If you are caught outside, stay away from—and lower than—high or conductive objects, including trees.
- Thunderstorm rains cause flash floods. Be careful where you take shelter.
- Winds. Very strong straight-line winds can cause great damage.
- Large hail. A rare killer, but very damaging.

Courtesy of the U.S. Department of Commerce, National Oceanic and Atmospheric Administration, Public Affairs Office

"The Tornado" portrays "how we used to beat it to the cellar before the storm hit." Painting by John Stewart Curry. (*New York Public Library*)

IV. FLOODS

Floods are a common occurrence all over the world, and few areas of the United States are totally immune to them. By definition, a flood is an accumulation of water that is too great to be disposed of by the normal means. While the absolute quantity of rainfall or snowmelt is a significant factor, soil concentration and topography are frequently contributing influences. Hence, flooding may be due to poor drainage, frozen soil unable to absorb excessive rainfall, narrow rivers and streams unable to contain billions of gallons of water cascading down hillsides, urban drainage problems, and hurricane storm surges. Flooding, of course, was mentioned in Biblical times when we are informed that it rained for forty days and nights, and, while nothing so extreme has occurred in modern times (and such an event is unlikely, at best), Noah at least had adequate notice to build himself an ark. Victims of contemporary flooding should have it so well.

Even Calama, Chile, situated in the rain shadow desert of Atacama, and which had enjoyed the reputation of being the driest place in the world, was subject to a disastrous flood recently. Having had virtually no rain for nearly 400 years, this town was the subject of torrential downpours on February 10, 1972, which caused catastrophic floods and landslides. Here, the novelty of the event and its effects on the soil undoubtedly contributed to the disaster, while in other areas of the world, such as the fertile valley along the Tigris and Euphrates rivers or along the midwestern Mississippi River, a year without some flooding would in itself be something to write home about.

One uncomfortable fact about the history of lethal flooding is the contribution that man has made to his own demise. Faulty engineering contributed to thousands of deaths when a dam burst in Johnstown, Pennsylvania, in 1889 or in Williamsburg, Massachusetts, some fifteen years earlier. A disaster took place in Lake Okeechobe in the fall of 1928 due to a hastily constructed levee and claimed nearly 2,000 lives after a hurricane swept through and weakened the structure to the point of collapse.

Over the north country, the main flood season comes from early March to the last week of April as a result of warmer temperatures accompanied by heavy rains, falling on a deep snow cover. The great all-New England flood was an example of this phenomenon. A secondary season, along the Atlantic coastal states, is from late August until mid-October as moisture-laden tropical disturbances bring massive amounts of rainfall northward, as in the extreme floods of 1927, 1938, 1954, and 1955.

Some of the most lethal floods on record were the Big Louisville Flood of 1937, which killed upwards of 500 people. The combination of hurricanes Connie and Diane claimed 169 lives in the East, while hurricane Audrey claimed 534 lives in Texas and Louisiana in 1957. Two flooding events in southern California took a toll of more than one hundred in 1969. Similarly, in Minot, North Dakota, in 1976, many people lost their lives or were displaced due to flooding for the fourth season in seven years.

The worst flood in world history occurred in China in 1931 when approximately 3.7 million people were drowned when the Yellow River overflowed its banks. Similarly, a cyclone accompanied by an earthquake in 1741 brought a storm surge ashore in the Bay of Bengal, India, nearly forty feet above mean low tide, killing over 300,000 people.

Perhaps the best advice regarding flooding is to be aware of the history of such events in your community and be prepared to evacuate, if and when necessary, flood-prone areas. An appendix appears at the end of this chapter.

21. A DISASTROUS FLOOD IN MASSACHUSETTS

"HAPPILY," REPORTED THE NEW YORK TIMES IN THEIR MORNING EDITION OF SUNDAY, May 17, 1874, "it is seldom we are called upon to record a calamity so sudden and disastrous as that which yesterday morning overtook three prosperous villages of Massachusetts." Rain had been falling steadily in the days preceding the event, and the ill-fated reservoir overlooking the town of Williamsburg was filled to capacity and leaks sprung from the bottom portion of its bulwarks. In the manufacturing valley below, 3,000 residents were eating breakfast or preparing to go to work when billions of gallons of water were set free.

The villages in Hampshire County, Massachusetts, about ninety-five miles to the west of Boston and twenty miles from Springfield, were dedicated to the manufacture of iron castings, edged tools, carriages, gold pens, woolen goods, and other articles, and were dependent upon the reservoir for power. The structure was constructed in the mid-1860s, during the Civil War and was filled in 1867, as the Mill River was undependable as a source of water. The Goshen Reservoir, the cause of the calamity, covered about 150 acres of ground, about a mile square and forty feet deep on an average. It was deemed tolerably secure, although it was given to leaks and had caused apprehensions of disaster in the past.

George Cheney, the keeper of the dam, lived with his wife and children and father at the top of the hill, 225 feet above the village of Williamsburg. The watchman made a practice of inspecting the structure every morning upon rising at 6 A.M. This Saturday morning was no different, although he did notice some minor leaks at the bottom of the dam. Being nothing out of the ordinary, he joined his family for breakfast. Suddenly his father, looking out the eastern window shouted, "For God's sake, George, look there." A section about forty feet in length at the bottom of the reservoir and just beyond the gate was shooting down the stream. With a single eye towards discharging his duty, and no thought of the peril involved, he rushed down to the gate and opened the valves fully, in hopes that it might afford some relief to the impending disaster. When the breach widened, he headed to the stable, mounting his horse, galloped full speed down the hill to the residence of the owner of the dam, a certain Mr. O.G. Spellman. When he arrived, he was met with incredulity when he intoned, "The dam is breaking." The two of them had discussed the structure's condition only the night before, and considered it safe. The roar of the thundering waters convinced Spellman of the impending trouble, and they dispatched a second horse to the town of Williamsburg, which lay about two miles downstream.

Precious moments were wasted, and when the messenger arrived at a factory in southeast Williamsburg he was greeted coldly by the foreman. "The flood is coming," replied Cheney when the foreman inquired as to what his business was. Cheney had been laughed at in the past, and the man replied, "Oh, yes, it will be here in about three days." He had barely uttered these words when a wall of water about fifty feet high and bearing splintered household goods rounded the bend towards the town. An alarm went out and 249 of the 250 hands were saved. Only the foreman, who had returned for a pair of shoes, perished in the inundation. The roar of the water was louder and more appalling than thunder, and rooted listeners to the spot. Before their very eyes, a house was borne on the summit of the torrent, with a woman and child looking through the windows and screaming for help. Dwelling after dwelling in the hapless town was carried away.

Never before in New England had an accident of a similar nature been attended with such sad and fatal consequences. Within a half an hour, the wall of water swept through the villages of Williamsburg, Skinnersville, where it wiped out a silk factory, Leeds, Haydenville, and Florence. Only when flood reached Northampton, nearly fifteen miles downstream, had it spent most of its fury. Yet still the water rushed out over the meadows and around the base of houses driving out the residents, compelling them to seek refuge on the higher bluffs surrounding them. In Northampton, a railroad span 100 feet across was torn away, and the people of that town turned out as a single body and lined the streets near the river, watching its destructive effects with fearful anticipation. When it became evident that the worst had been experienced, they turned their attention to the stricken valley, "anxious to afford all the aid in their power."

Upstream, in Florence, the rescuers found that that town had been spared by its topography. Cook's Bluff, a natural ridge rising one hundred feet above the surrounding area, had successfully turned aside the furious current. The meadows beyond distributed and dispersed the flood and rendered it powerless to injure. But in Leeds, a factory town of about 5,000 people, it was a different story. The bend in the river only served to intensify the deluge, and a woman, child in her arms, stood on a hill overlooking that city and could only say, "My John and my boy lie under those timbers." So great was everybody's personal interest in the wreck before them that few found time or interest to comfort the poor creature, for she was only one of hundreds stricken.

As the great bell in the old Leeds church tolled solemnly, calling people together to search for the missing, notices were posted in conspicuous places in town reading, "All able-bodied men are hereby summoned to the relief of the people of the Mill River Valley." Men on horseback and buggies struck out armed with axes, crowbars and spades and worked away at the mounds of driftwood and wreckage, in quest of bodies. In Leeds, the heroism of Myron Day came to light as a modern day Paul Revere. A little before eight in the morning, the people of the town were attracted to this young lad, riding through at top speed on horseback shouting wildly, "The reservoir is broken, save yourselves, for the flood is at hand." People looked northward toward the Goshen Dam and saw what seemed to be the crest of an enormous pile of moving wood and knew immediately what it was. In less than three minutes, a great column of water laden with every conceivable household item plunged into the valley of Leeds. Everything in its path was smashed to atoms and young Myron Day barely escaped with his life, after saving hundreds of people in that ill-fated village.

When it was all over, nothing but a sewing machine embedded in the mud revealed that a house had once stood there, as the water had stood fifty feet deep in the pouring rain. All through the day and night, people strained their eyes to catch a glimpse of missing ones who might be clinging, perhaps, to part of the wreckage. Others hunted up and down through the crowds, and called out across the flood to the opposite banks, inquiring whether their loved ones were on either side. At length, the doubts of some as to the fate of relative or friend were set to rest by the discovery of bodies further downstream in Florence or Northampton. In Northampton, church bells broke the stillness and numbing rain, and the dead were buried unostentatiously on Sunday. People came a long distance in market wagons for their dead and mourners followed behind on foot.

One of the most remarkable escapes was that of Mrs. Mary Harding, of Leeds, who was at work in a silk factory when the flood neared. She led the whole company out of the building and considered crossing the river to the safety of higher ground. Shouts cried out, "Run across the bridge!" and "Come back...don't go over." It was too late. Turning back was just as dangerous as pressing forward. She ran, as perhaps a woman had never run before and reached the further shore with six others when an immense mass of debris struck the bridge. Many of her companions met an agonizing death at this time but she headed up a steep embankment towards Ross's General Store rather than seeking the temporary shelter of an ill-fated boarding house. Just as she reached the steep steps of the store, the water rose sufficiently to knock her off her feet, and it would have swept her away had it not been for the courage shown by two men nearby. Of the thirteen people who endeavored to cross the bridge, she was the only one to escape with her life.

A particularly poignant story was told in the village of Williamsburg, where sixty-year-old Ira Bryant had walked his favorite dog, a St. Bernard, into town as a daily ritual. Mr. Bryant was crushed by the torrent as it first entered the village and lay buried several feet beneath the mud. A search party had tried and failed to locate his remains, but his best friend pawed furiously in the sludge and attracted a great deal of attention. Upon inspection, the lower limbs of a body were revealed, and as the face was revealed the dog seemed overjoyed. However, when a cloth was wrapped around the rigid form and carried away, the noble creature seemed bowed with grief and sulked behind the crowd. A diligent search had been made for the man, but it seemed hardly possible that any human being could have found the imbedded corpse.

The town hall in Williamsburg served as a morgue where the drowned were bandaged and stretched and put into their coffins. The small room was filled with weeping men and women and children. Mothers sat by the recovered bodies of their children, and children by the remains of their parents. Nothing could be done to scour the earth for additional victims, as the waters were yet high and people were paralyzed with grief and found sufficient occupation in caring for their bereaved neighbors.

In a couple of days, a move was afoot to rebuild the damaged factories, and all hands were employed in the task of cleaning up the debris. After any tragedy, there are always bands of thieves who prey upon the helpless, and in this case most made off with scattered valuables, as there was no one around to stop them. In the ensuing days, Mr. O.G. Spellman, the overseer of the ill-fated reservoir, incurred a large share of the blame for the disaster due to this foolish detention of George Cheney that Saturday morning and the common perception that the reservoir was never safe from the start. A group of vigilantes plotted to drag him from his home and drag him into the woods where they would lynch him, but it was foiled when a former employee who had heard of the scheme informed him only an hour before it was to take place.

In retrospect, it is true that the reservoir was constructed shabbily, for at the base, which was regarded to be twelve feet thick, the remaining stone revealed that it was just over five feet in thickness and tapered to a paper thin two feet at the top. Similar to the Johnstown tragedy just fifteen years later, it appears that the proprietors cut corners in financing the venture and a good solid rainy period was all it needed to nudge it over the top. Needlessly, three villages in the western valleys of Massachusetts were swept away on a gray rainy day in May of 1874. The scars that those communities will have to bear will live on for generations, for no one living there now could be unaware of the awful tragedy that befell their ancestors more than a hundred years ago.

22.
JOHNSTOWN, 1889

THE NOTION OF A CALAMITOUS FLOOD IN JOHNSTOWN, PENNSYLVANIA, WAS NOT SOMEthing new in the spring of 1889. In fact, residents were accustomed to frequent flooding, as though it were an annual event similar to what the ancients experienced with the Tigris and Euphrates. Eighteen miles up the Conemaugh Creek, beyond the workingmen's villages of South Fork and Mineral Point, was Conemaugh Lake. The lake was actually part of a canal system and a natural lake in among the hills of Pennsylvania, some 300 feet in altitude above the city of Johnstown. A club called the South Fork Fishing and Hunting Club got use of the property and renovated the dam to a point where its walls stood 100 feet high and 900 feet across. This greatly increased the size of the lake to a volume and shape roughly the size of Manhattan, about one mile wide and five miles long. And it greatly increased the risks to the 55,000 residents in the valley below.

This work was completed over a two-year period between 1879 and 1881. It was considered amply secure, for the members of the club actually used the top of the dam as a driveway. Usually the water level of the lake was fifteen feet below the top of the dam, yet while the work was going on in 1881, a sudden rise in the lake's capacity sent workers scurrying to prevent a complete washout. A vigorous effort was made to stop this construction, for people were afraid of the dam, and well they might be when one contemplates the wall of water behind the 100-foot-high edifice. The foundation of the dam in the spring of 1888 was considered shaky, and many leaks were reported from time to time. People wondered why it had not been strengthened, and there was probably no one downstream in Cambria County who had not feared the awful consequences of a failure at one time or another.

Rumors often circulated among the citizens of Johnstown that danger was imminent, and a few would always head for the nearby hills only to return and be branded cowardly. It seemed the fear of public scorn was greater than the fear of the flood itself, so it is no wonder that when the wolf came down upon the fold, most ignored the cries of "Wolf!" On Friday, May 31, 1889, they say the avalanche of water from the lake pounced upon the town of Jamestown so suddenly that there was no possibility of escape. It was on that day that the angel of death spread his wings over the fated valley, unseen and unknown. Almost, that is. A certain Herbert Webber, employed as a guard at the boating club, had protested the condition of the dam to his employers before and was reprimanded

Debris of the Pennsylvania Railroad stone bridge. (*Library of Congress*)

under the threat of dismissal. He took his case to the mayor of Johnstown a month earlier and was assured that an engineer would be dispatched to inspect the situation. This never happened. On the morning of the thirty-first, Mr. Webber was on patrol about a mile away from the dam when he noticed a sudden drop in the height of the lake. This was peculiar, for in the past thirty-six hours, rain had been falling in torrents and the mountain streams feeding the lake were swollen and running fast, and the lake itself had been rising at the rate of ten inches an hour. It had been close to spilling over.

Just three days before he had noticed the water forcing itself out of the interstices of the masonry so that the front of the dam resembled a huge watering can, with jets which squirted out thirty feet horizontally. Shortly after one in the afternoon, he reached the dam and saw water welling out from beneath the foundation. Utterly helpless and awe-struck, he was compelled to stand there and watch was to be the most disastrous flood of the continent up until that time. In another, somewhat contradictory account, it was reported that a crew of twenty-one workers had been laboring furiously on the dam all morning, and when they saw that their efforts were fruitless, a sentinel was sent out by horseback to warn the people in the neighboring town of South

An artist's conception of the Johnstown flood. This first appeared in *Harper's Weekly* on June 15, 1889. (*New York Public Library*)

Fork. From there, word was spread by telegraph to the villages downstream, and one observer said that the message was received in Johnstown about noon and posted in the center of town where many mingled in disbelief. After all, they had heard it a hundred times, and the warning had proved so useless in the past that little attention was paid to it at this time. That was three hours before the flood hit, and men came home from work early to move their families and provisions to the second floor to ride this thing out. It was too bad that the dam had not been repaired after the spring freshets of 1888.

The last notice to be sounded was from a train whistle from an engineer who watched in horror when the dam gave in. Witnesses say the dam did not burst as most dams do but simply moved away. It produced a wall of water thirty to forty feet high which moved down the slopes with the swiftness of a mountain lion. A group of seventy-five to one hundred people had gathered on a nearby ridge to see what was going on. When the flood waters approached, they were seen to throw up their hands in dread before they were washed to their deaths. Thirty-three locomotives were washed away from a nearby roundhouse, and one, which was occupied, was heard to whistle by the steadfast engineer eight miles below and twenty feet beneath the surging waters of Johnstown.

At 3 P.M. the telegraph operator in South Fork was gingerly giving a blow-by-blow description of the flood and excused herself to the upstairs office when the waters reached three feet. At 3:07 P.M. communications were lost and so was the community of South Fork. Houses, factories, and bridges, as well as their occupants were swept away immediately in an avalanche of destruction. The next town in its way, Mineral Point, was obliterated in a tangle of timbers and human souls. By 3:30 P.M., the flood reached Johnstown. By 3:40 P.M., Johnstown was no more. One boy, who was rescued ten miles downstream, related the following story:

> Shortly after 3:30, there was the noise of roaring waters and screaming people. We looked out the door and saw people running. My father told us to never mind as the waters would not rise further. But soon we saw houses being swept away and we ran up to the floor above...in my fright, I jumped on the bed. It was an old fashioned one with heavy posts. The water kept rising and soon my bed was afloat...the air in the room grew close and I was pressed against the ceiling. At last the posts pushed through the plaster and a section of roof gave way...then suddenly, I found myself on the roof and was being carried away. I would see people, hear shrieks and then they would disappear.

So strong was the public mentality against the possibility of disaster, this boy's father believed it until the very end.

Men women and children were carried away in the flood, crying frantically, but their screams availed them nothing. Rescue was impossible. Husbands were swept past their wives, and children were born along so rapidly to certain death, even before the eyes of their terrorized parents.

One of the earliest dispatches to reach the outside world from Cambria County read, "Johnstown is annihilated." It wasn't far from the truth. "Johnstown is completely submerged and the loss of life is inestimable, houses are going down the river by the dozens and people can be seen clinging to the roofs." The Juniata River, normally as placid as a creek, was over thirty feet deep when it roared through nearby Hollidaysburg. Clearfield was under four feet of water and the town of Dubois was completely destroyed. Further down the river in Bolivar, one witness related, "We knew nothing of the disaster until we noticed the river slowly rising and then more rapidly. News reached us from Johnstown that the dam at South Fork had burst. Within three hours, the water in the river rose at least twenty feet. Shortly before 6P.M., the ruins of houses, beds, barrels, and kegs came floating past the bridges without interruption. Then it began to lessen, and with night coming suddenly upon us, we could see no more."

To add to the terror that evening, the sky was illuminated by the light from hundreds of burning houses that were piled up fifty feet high in a jam created by the debris of the old stone railway bridge against the western shores of Johnstown. Some people were wedged into this pile-up so tightly that it was necessary to amputate their legs to free them. Contemporary accounts differ on the number of people who died in this conflagration, but it seems certain that there were at least 700 and possibly as many as 2,000. Two days later, the fire was burning as fiercely as ever. In the inextricable confusion of houses, locomotives, timbers, and people, the work of counting the dead was so appalling that its horrors can be better imagined than described.

The only building in the business district to survive, Alma Hall, housed nearly 200 people that fateful night where they took refuge on the upper floors. A local attorney described the scene:

> No lights were allowed and the whole night was spent in darkness. The sick were cared for...the scenes were most agonizing. Shrieks, sobs and moans pierced the gloomy darkness. The crying of children mingled with the suppressed sobs of women...no one slept during the long dark night. Many knelt for hours in prayer, their suppli-

View of the Conemaugh River after the disastrous flood. (*Library of Congress*)

cations mingling with the roar of the water and shrieks of the dying in the surrounding houses. In all this misery, two women gave premature birth to children.

When dawn broke the following day, the scenes of devastation were everywhere. One woman, identified as Mrs. Fenn, described her ordeal to a reporter. She recalled that as her house was crumbling, she raised a window and, one by one, placed her children on driftwood passing in the night. "As I liberated the last one, my sweet little boy, he looked at me and said, 'Mamma, you always told me that the Lord would care for me; will He look after me now?' I saw him drift away with his loving face turned toward me and with a prayer on my lips, he passed from sight forever. The next moment, the roof crumbled in...I have lost everything on earth now but my life, and I will return to my old Virginia home and lay me down for my last great sleep."

Another woman, with hair as black as a raven's, poked around through the hundreds of recovered bodies, carefully lifting each sheet and then replacing it until she found her sister. She spoke softly, "...and all her beautiful hair all matted, and her sweet face so bruised and stained with mud and water." As she spoke, tears came to the eyes of the biggest and sturdiest of men.

As always, at the scene of a great disaster, stories circulated concerning miraculous escapes. Just before the forty-foot wave crushed Johnstown, an alert trainman pulled away from the Johnstown station at breakneck speed, crossing the bridge which spanned the Juniata River to safety, moments before the bridge was swept away. Further down the river in Cambria City, the Church of the Immaculate Conception was holding a service when it was overwhelmed by the flood. Only one object in that city was left standing—a three foot statue of the Blessed Virgin that had been decorated and adorned with flowers, wreaths, and a lace veil. Survivors claimed that the statue was spared all contact with the water, which had risen to a height of fifteen feet, but the statue was "as unsullied as the day it was made." Nothing short of a miracle saved the life of a five-month-old baby boy, whose cradle was found floating on a piece of driftwood in the Allegheny river near Pittsburgh—some ninety miles away from the scene of the deluge.

Unfortunately, inhuman degradation frequently follows a disaster of this scope, and the first trains to arrive from the west did not contain relief but bands of thieves whose sole interest was to prey upon the dead, removing cash and jewelry from the dead. There were instances of the victims' fingers being cut off, to facilitate their haste and their greed. The townspeople fought back, in some cases, and there were several stories of on-the-spot lynchings while others were driven by the angry crowds to a certain death, in the still surging waters which engulfed Johnstown.

Relief, however, did arrive in a few days, and it was through the noble efforts of Miss Clara Barton and her Red Cross organization that makeshift hospitals were erected in the hills of Stoney Creek Village to battle the pestilence that frequently survives such destruction.

An exact tally of the dead will never be known, but contemporary accounts ranged anywhere from 7,500 to as many as 15,000 citizens in the villages within a twenty-mile range of the lake. Modern opinion rests upon the figure of 3,100, but this is entirely too low, for as of the fifth of June, 3,500 bodies were accounted for in the city of Florence alone. Needless to say, arriving at an exact casualty figure was not one of the priorities of the rescue operation; the health and welfare of the living was.

Early accounts of the disaster blamed some meteorological phenomenon such as a waterspout striking the dam, but this too can be discredited. Rather, it was the decrepit condition of the dam, the callous indifference of its caretakers and, to a degree, the hardy abandon with which the people of this valley regarded the lake. One no sooner leaves the catastrophe at Johnstown when the front pages reveal that the entire city of Seattle, Washington, was burned to the ground in a fire caused by a bottle of turpentine. So it goes.

23. THE GREAT MISSISSIPPI FLOOD OF 1927

1927. Babe Ruth was already well on his way to establishing his monumental home-run record which was not to be eclipsed until Roger Maris came along in 1961, and then only with an asterisk. 1927. The era of the first trans-Atlantic flights, featuring races from New York to Paris with names such as Lindbergh and Byrd, destined to be indelibly printed on the minds of three generations of Americans. 1927. The year of American heroes and the year of the most widespread, terrifying flood in the history of this country.

Bad weather is routine in the Midwest in the spring, and the Mississippi floods to some degree nearly every year. Violent weather visited the middle of the country in the form of tornadoes, cutting a two-mile-wide path through Texas, killing 150 people on the twelfth of April. A week later, thirty-one citizens of the city of Springfield, Illinois, perished when a tornado roared through their town, with additional damage in Missouri and Nebraska from a heavy hailstorm. Nearly fourteen inches of rain fell over the area in the first two weeks of April, and by the eighteenth, nearly twenty percent of the continental United States lay under the floodwaters of the Mississippi River.

By the twentieth of April, more than 100,000 people from Cairo, Illinois, to Helena, Arkansas, had been evacuated as one community after another saw their valiant efforts go for naught as the relentless forces of the Mississippi crumbled their bulwarks. The crushing power of the Mississippi riding higher against the great dikes which partially confine it to its course caused apprehension from Illinois to the Gulf of Mexico. Millions of acres were inundated due to backwater breaks which overflowed tributaries, spilling out yellow waters one hundred miles long and one hundred miles wide.

Refugee camps were set up apparently out of the path of danger, dotting the Mississippi, and outbreaks of measles, mumps, and whooping cough were a tragic feature of these centers before the centers themselves were flooded in Arkansas between the White and Cache rivers. When the White River overflowed its banks near Clarendon, a village of 3,000, the town was transformed into a foaming whirlpool where "men and animals and swaying houses were tossed about in disorder", marooning hundreds of refugees on Tom's Hill. Houses, animals, and rivercraft were washed down the main street of the business section, and many without telephone service were unaware that the levee did not hold until it was too late. In Arkansas, between Little Rock and Pine Bluff, refugees were seen clinging to the tops of church steeples and in trees to escape

Artificial crevasse formed at Coernorvon, Louisiana, below New Orleans, on April 30, 1927. (*Library of Congress*)

the rising tides. At the time, the river in Memphis was 44.7 feet above flood stage, and in Helena it was an astonishing 56.5 feet.

Meanwhile, the rains had begun anew, bringing a total of 4.5 inches to Memphis, seven inches more in Little Rock, and a staggering ten inches to Benton. Pestilence was now riding on the waves of the flood, and it was finally the St. Francis River that broke through to invade a major refugee camp in Little Rock. Rescued by steamboat, motorboat, and barge, thousands whose misery was compounded by cloudburst, high winds, and rising waters joined the estimated 50,000 others at Red Cross shelters further "inland." Further downstream, men were arrested and, in some cases, shot for dynamiting the levees on one side to ease the pressures on the other, presumably to save their homesteads.

The twenty-first of April brought more rain and freezing temperatures as the floods continued to rush forward with unexpected intensity. The wall of the flood waters broke through a breach at Stops Landing, Mississippi, and fire engines screamed alarm, warning residents that the break had come. Workers were handicapped because all the dirt and sandbags had to be hauled over roads partly under water. At Knowlton Landing, the boat launch *Pelican* was pulled through a levee break; it capsized, drowning all of the eighteen refugees aboard just before it reached the safety of the steamer *Wabash*, which was passing up the river.

An indication of the strength of the stream was that the Arkansas River

Flood waters from the Mississippi roam through the town of Melville, Louisiana. (*Library of Congress*)

had washed away two spans of the Missouri Pacific railroad bridge at Little Rock, carrying with them a string of freight cars loaded with rocks and coal that had been run into the bridge to anchor it against the rising floodwaters. Earlier, the bridge shook so violently that coal was set ablaze from the friction. A few moments later, the structure toppled into the water, never to be seen again.

Day after day came reports of towns further south that had been overwhelmed by the advancing waters. In preparation, levees had been fortified with sandbags to stem the rushing tide, and day after day new breaches occurred, creating new ruins. Greenville, Mississippi, was swarming with 10,000 refugees, and it too was swamped under ten to fifteen feet of water. Riverboats were desperately trying to reach the victims, and it was advised that "hundreds had taken to the tops of levees and housetops and unless they are reached in short order, they undoubtedly will be drowned." The Mississippi flood was declared the worst in the state's history, and the "whims of the surging torrent are hourly becoming more perplexing to those who profess to know the antics of the mighty stream when on a rampage."

By the twenty-fourth of the month, it was reported that one of the "greatest peacetime battles in all of history" was being fought over an inland sea covering thousands of square miles, with "every conceivable method of rescue" being pressed into operation. Back in Helena, where 3,000 persons had been saved earlier by the *Wabash*, cries of help came from the top of the levee where people stood stretched out for three miles. Fliers risked their lives in saving them from the raging waters, as it was not always possible for boats to reach those in distress. Parachutes, of course, could not be used; they would have been worse than useless.

Fugitives from the water-infested town of Greenville were rescued by steamer. Upon their arrival in Vicksburg, they were seen tripping and staggering down the gangplank for their first food and shelter in nearly a week. Many of them, as they fell into the embrace of anxious relatives, had trembling hands and darting eyes. One old black man, apparently without

The Mississippi flowing through the Bayou de Calaise, Louisiana. (*New York Public Library*)

family or friends, commented to a reporter, "I don't know where deys goin', but I's gwina foller um."

The breaks in the upper reaches of the Mississippi did not diminish the great volume of water that was surging south. In fact, most of the water made perilous side trips before rejoining the main river that was calculated to be flowing past the doomed towns at a rate of 3,000,000 cubic feet a second and now covered over six million acres of land. On both sides of the Mississippi, water was thundering past, while the marooned awaited the arrival of craft of mercy and the relief of the American Red Cross. It was to be thirty to forty days before the waters would recede, and some never returned to their homes, because they had no homes to return to.

All the while, New Orleans was laboring as never before to save great parts of the gulf metropolis from the crest of the flood. The Mississippi was no longer a river but a great lake, extending from St. Louis all the way to New Orleans, and the entire state of Arkansas was no longer visible from aircraft. It would have been interesting to see a satellite photo of the region, had that been available at the time. What was visible were the points of high ground, for

Barge loaded with flood refugees. (*Library of Congress*)

wherever they stood, refugee camps were organized with army tents pitched. The army and Red Cross had rushed trainloads of tents, field kitchens, blankets, and medical supplies to meet every possible emergency in the path of the flood. The only other sight visible from the air was the tops of giant oaks and willows that looked like an unmowed lawn on top of the waters. Motorboats sped between treetops on missions of mercy where thousands awaited rescue. The volume of water passing Memphis was ten times greater than that pouring over Niagara Falls when the Niagara is at maximum flood stage. These were conditions which Americans who had never seen a flood like this could not possibly visualize.

Secretary of State Herbert Hoover was put in charge of directing the relief operations, and it was his decision, with the aid of the Army Corps of Engineers, to blow up the Poydras Levee near New Orleans to save that city. Thirteen miles south of New Orleans, more than 10,000 people were asked to leave their homes for good, like a condemned Eden, for it was guessed that after the Mississippi roared out into the Gulf of Mexico, not even the soil would be left. A 1500-foot break in the levee was planned, and movie cameras were prepared to film the spectacle for the nation to see. On Friday, the twenty-ninth of April, the "big explosion" took place, thousands of tons of dynamite being detonated. After six blasts, a relatively small breach was established and the mighty Mississippi roared through, taking houses and trees with it. The roar could be heard for miles around as the waters of the Mississippi merged with Lake Lery, channeling their forces into Breton Sound and finally into the gulf. It took nearly two weeks to create a crevice nearly one thousand feet long, and this has served as a permanent outlet for the Mississippi River to relieve New Orleans in future floods.

The state of flood was not over for quite some time. The river, or at least the crest of the river, moved south at only forty miles per day, creating the greatest area of devastation and destitution the nation has ever known. Everywhere along the river, people constructed fortifications, the river winning every phase of the battle. One town after another disappeared under fifteen feet of water, and it is estimated that the Red Cross provided for nearly three-quarters of a million refugees. When the Bayou des Glaises Levee tottered and crumbled, it caught a sleeping town of thousands unaware, and only the shots from National Guard rifles and the clanging of bells averted the nation's largest casualty list.

The first two weeks in May compounded the devastation of the previous month, as tornado outbreaks swept through the Midwest and South, killing hundreds more and injuring thousands. It was the arrival of "Lucky" Lindbergh in Paris, 33½ hours after his solo departure from New York, that finally moved the weather of the spring of 1927 off the front pages. America was ready for a hero of this stature. After all, Babe Ruth was soon to be struck out by seventeen-year-old left-hander Jackie Mitchell, a girl.

24.
THE SPRING FRESHETS OF 1936

After an unusually hard winter season of 1935-36, residents of the United States from the Ohio Valley east to the Atlantic and northward to Maine were heartened when the weather finally broke on the tenth of March. Snow persisted and lay in an unusually deep cover throughout that winter season, and on that pleasant Sunday temperatures soared well into the sixties in a broad swath up and down the East Coast and back to the west as far as Chicago and St. Louis. It appeared that winter was receding into the northern hinterlands, and the ice and snow accumulation that had been so companionable over the past three or four months was finally going to melt. That assessment was eminently correct, but any notion that the sweet early days of spring lay just around the corner was just poet's fancy.

The eleven-day period from the eleventh to the twenty-first was to spawn three devastating rainstorms, with barely time for the sun to shine between them, and the consequence of so much rainfall on top of the softened snow was to produce the worst floods in the industrial northeast of the century. A slow-moving coastal storm from the southeastern states tracked northeastward to the Chesapeake Bay region on the eleventh, unloading torrents of rain on the white countryside as it went. Only hours after the first thick raindrops splattered down in the Mid-Atlantic and Northeast, rivers began to swell over their banks, creating huge ice jams and sending their swirling waters through large areas of the Northeast. Highways were blocked and water invaded scores of communities where runoff was at a maximum due to the frozen, saturated condition of the ground. Northern New Jersey was particularly hard hit, as many residents had to be rescued by police boats groping around in the dense fog with lanterns on the night of March 12, a day which already lived in infamy as the commencement of the great Blizzard of 1888.

The storm pressed northward into the St. Lawrence Valley on the thirteenth unloading up to seven inches of rain on the hill country in New Hampshire, Vermont, and Maine, and its impact on the average of three to five feet of snow set into motion a series of freshets which washed away bridges, highways, and railroad lines. Men labored furiously along major river systems, piling load after load of sandbags to turn back the raging waters of the Connecticut River. In Lawrence, Massachusetts, the river reached a record of 43.9 feet, higher than the disastrous flood in that region of 1927. Yet, still the rain poured down the mountain slopes, gathering and melting the deep snows in its wake, and rivers throughout New England continued to swell long after the rains abated.

The worst of the danger appeared to pass by the fifteenth of the month, with the regions in central and northern Pennsylvania, along the Juniata and Susquehanna river basins still experiencing high waters. The weather improved so much, in fact, that sightseers thronged to the stricken areas and, in particular, in Passaic Falls, New Jersey, where thousands gathered to watch the swollen waters plunge 100 feet into the chasm. Atlantic City, not seriously affected by the turn in events, entertained tens of thousands of visitors along the Boardwalk as people turned out in an Easter-like promenade. The temperature soared to sixty degrees once again.

All would have been well had new rains from the south not commenced on the sixteenth. While Buffalo, New York, reported a new snowfall of twenty-four inches, areas well to the south and east were pelted by drenching cold rains and gale-force winds of forty-five miles an hour. Once again, Johnstown, Pennsylvania, was in the news. Floods exceeding fourteen feet were reported in the streets of what has become known as Hard Luck City. Tens of thousands of people, recalling the fateful flood of 1889, fled the city, while many others were marooned in buildings where muddy waters lapped up to second storey windows. Graphic details flowed in from Johnstown to the flooded world outside. The rains had inundated the city by 1:30 P.M. on the seventeenth, and the swirling brown currents raced through the streets with such swiftness that rescue work was rendered virtually impossible in a city without lights. By 2:30 P.M., those who remained behind were forced to take shelter in the upper floors of the town's buildings. Walking in the city had become impossible, and radio networks set up monitoring stations around Johnstown.

In what has to be one of the most dramatic and brief bits of radio reportage ever heard in an emergency, news flashed across scores of participating stations, interrupting them without ceremony, that the giant dam above Johnstown had burst as it had in 1889. "The dam has burst, everyone run for your lives," echoed throughout NBC studios across the country. Radio engineers shouted that they had to leave, while New York monitors begged them to stay at their posts and remain. They did, for two and a half minutes. One reporter, who was hastily donning his shoes, had the last word from Johnstown. "We are ready. We have to get out of here. We are leaving," he blurted into the microphone, breathlessly. "The dam has burst. Hello New York! They have ordered us out. We are leaving the building. We're heading for the hills!"

The residents of Johnstown were again huddled on the rain-soaked Conemaugh Hills above the city, battered by flood waters from the Conemaugh River and Stony Creek. As it turned out, the huge Quenehoming Dam had not broken but a smaller one at Wilmore, nine miles up the river, had. However, the flood waters were already so high that the new breach caused no appreciable change in the flood condition, but rumors continued to fly among the sizable knots of people gathered upon the hillside looking down over the wreckage. At least, in this instance, they had erred prudently. Several employees of the Bell Telephone Company chose to remain in the city and continued to provide a blow-by-blow description of the rising flood waters, until, at last, all power in the city failed. Among the refugees, there was an outbreak of scarlet fever and diphtheria, and a dance pavilion high on Westmont Hill became a cheerless

This young man has found a unique perch for surveying the scene in Hartford, Connecticut, following the March floods of 1936. (*New York Public Library*)

makeshift hospital for 658 sick and destitute persons, including 253 children. Three babies were born in a span of forty-eight hours, orphans of the flood. Viewing the town from the hill in the daylight, one could see huge trees lying on rooftops, telephone poles blocking bridges, and automobiles piled up against buildings. The high-water mark had equaled that of 1889.

Further west, in the vicinity of Pittsburgh, rain and wet snow pelted the Steel City, and the usual veil of smoke that shrouds the city had vanished as its three major power plants had failed in the rising tide. The flooding was most severe at the junction of the Monongahela, Allegheny, and Ohio rivers and actually exceeded, by eight feet, the disastrous high-water mark set in 1907. Waters rose to an all-time high of forty-eight inches above mean low water and marooned tens of thousands in downtown office buildings. Like Johnstown, the waters rose dramatically and with little notice and submerged the "Golden Triangle," the heart of the business district, under more than ten feet of water. A report from suburban McKees Rock disclosed that 500 men, women, and children were stranded in the upper floor of the Blair School, and frantic calls for aid went out as they feared that the building might crumble any minute.

The city was plunged into darkness that was to persist for a week, and when darkness descended on the eighteenth, it was a night of terror for thousands, as numerous fires and explosions heightened this apprehension. Several victims of the raging flood were rescued from rooftops and treetops, and a story of a most daring rescue was told as three teenagers had capsized their motorboat and had to swim for their lives to the nearest treetop. Two firemen launched a rowboat in an effort to rescue the boys and met a similar fate as did the second rescue craft. Finally, a third boat succeeded in saving the lives of the seven, just before the tree went under. Misery loves company, and by the twentieth, the heavy

At the Red Cross Station in New Etna, Pennsylvania, refugees receive much-needed food as shown in this photo originally seen in the *New York Herald Tribune*. (*New York Public Library*)

rains had turned to snow, two feet of it in some sections, but the hidden blessing was that it stemmed the fires and disease in the prostrated cities and towns of western Pennsylvania, southward to Wheeling, West Virginia, and Cumberland, Maryland.

An amusing story emerged from an island near Port Jervis, New York, where a man known as the "island hermit," Bill Marvin, refused rescue attempts made by a couple of policemen in a motorboat. He set his four dogs upon them and, after enduring several bites to their legs, they killed the animals. Then the old man came after them with an axe, but was finally subdued, "put into the boat and is in jail tonight." One of his two cows was towed ashore while the other one swam to safety.

In the Wilkes-Barre/Scranton area of Pennsylvania, mines became underground rivers, filled by the inrush of flood waters. Thousands of miners lost their jobs for weeks before the mines could be flushed out. A radio station, WBRE, attempted to reunite separated families, without being sure that the marooned families had power to tune in to the radio with broadcasts like, "Will the Browns who live on Main St. please pay attention. Mary says that a boat will be there in a half hour." The Susquehanna itself was described as a massive lake, perhaps 444 miles long, stretching from Harrisburg to Lock Haven. Further south, the Potomac River reached 44.6 feet above mean low water, nearly six feet above the high-water mark of 1889 and an average width of a mile. Seven towns in the Potomac Valley near Washington, D.C., were totally deserted as a result of the high water, and the flood was visible to President Roosevelt from the White House.

Meanwhile, New England was experiencing floods from the ten-day downpour, perhaps hardest hit of all sections in the country due to the deep snow cover that began to slide down the mountains on the thirteenth. Ice jams were created on all major rivers, with perhaps the severest washout occurring at Hocksett, New Hampshire, where water eighteen to twenty feet deep rushed through the main streets. As Dr. David Ludlam points out in his fascinating *New England Weather Book*,

> The massive amounts of moisture would have created a serious flood problem even if they had fallen on bare ground, and the enormous amount of water locked in the snowbanks raised the March 1936 storms and floods to catastrophic levels. The snow cover in the White Mountains possessed a density of about 28%; that is, the 33 inches of snow on the ground at Pinkham Notch gave a water equivalent of about 8.00 inches. Adding the snowmelt to the total of 22.43 inches, rainfall during the two storms produced a total of 30 inches of available water to drain off the higher elevations.

The waters at Hartford, Connecticut, rose to a stage 8.6 feet higher than any previous flood level recorded in nearly three hundred years of observation, and a large section of the business district lay under water for several days as the Connecticut River backed up into the North Branch Park River, a tributary that flows through Connecticut's capitol. The raging Connecticut River forced the entire populace of Hadley and Sunderland to flee when it overflowed its banks and swept through those towns. Residents of Greenfield, Massachusetts, heard the dreaded ten blasts of their fire whistle shortly before 11 P.M. on the night of March eighteenth, and radio stations were broadcasting the news that the Vernon Dam had broken. Rumors flew through the town of Turner's Falls, the first below the dam, that a wall of water was approaching and that patients in the local hospital were in danger of being drowned in their beds. In fact, the dam was in trouble, but the all-night labors of a crew of volunteers saved the towns downstream, even though the water—filled with huge cakes of ice—was flowing about ten feet over the top of the dam.

More than 200 lives were claimed in the spring flood of 1936—115 in Pennsylvania alone—and more than $300,000,000 of damage was assessed. The unprecedented level of the water was compounded by a severe winter, but the amount of moisture dropped from the dark clouds in that period alone would have caused extreme problems. Even with modern engineering techniques, much of these circumstances could be repeated as the wary residents of Johnstown are well aware. The Northeast was truly overwhelmed by the powers of nature in the spring of 1936 and did not see anything like this disaster until the floods spawned by hurricanes Connie and Diane, nineteen years later.

25.
THE BIG THOMPSON CANYON RIVER FLOOD

"BELIEVE IT OR NOT" READ THE SIGN TO THE BILLBOARD AT THE ENTRANCE TO THE BIG Thompson River Canyon, at the foot of the Rocky Mountain State Park about fifty miles northwest of Denver. It was an advertisement for a Ripley's museum but believing was rather difficult for those who had seen the destruction wrought by one of the mightiest floods ever known in this scenic canyon in Colorado. Mobile homes and trailers were smashed into atoms against trees, while a steel water diversion pipe, fully nine feet in diameter and standing fifty feet above the usually tranquil Big Thompson River, had been torn from its trestle and lay jammed into a house a mile downstream.

The trouble began on Saturday night, July 31, 1976, just as Colorado was celebrating its statehood centennial. Thunderstorms overspread the area that evening and persisted for fully five hours without intermission. Billions of gallons of water cascaded down the steep slopes overlooking the river and poured into the narrow canyon. Perhaps as many as five thousand campers, who flooded the region in the summer for its tranquility and natural beauty, were in the area that night when the rains commenced.

The flood swept through between 8:00 and 8:30 P.M. and was all over within only twenty minutes. The unusual absence of wind that evening had allowed a thunderstorm to dump between thirteen and fifteen inches of rain into the narrow canyon as the Big Thompson River rose with a furious speed, trapping thousands. Lois Kelsey, a twenty-two-year-old camper, reported, "We thought we'd go out of town and get something different to eat. Boy! Was it different. Heavy rocks started falling around us, the rain got really bad and we decided to turn around...we were stopped by a river that just started coming into the canyon along with boulders, trees and mud." They pulled to the side of the road, scrambled out of the car and began climbing on their hands and knees to escape the rising waters. People from other cars began to follow suit. The group climbed one hundred feet to a cluster of overhanging rocks and sat under this natural shelter the rest of the night. Below they could see the headlights of their own car and those of several others being swept by the flood through the canyon. As the rising waters lapped beneath them, and boulders cascading above them, they sang songs like "Rocky Mountain High" to keep up their spirits. Instead of dining on lobster at the Black Canyon Inn, they made do with cashews, sunflower seeds, almonds, and grape juice.

In other areas along the twenty-five

mile stretch of canyon, things were no better. "It was the worst bedlam you've ever seen. I saw little babies being carried up the side of the mountain and old people climbing. These aren't small mountains, either. It was so cold that I thought we were going to die. It was just a bunch of water coming at you. You couldn't see it but you could hear it." Highway 34, a major thoroughfare entering the Rocky Mountain National Park, was almost entirely washed away. Another man reported that the canyon was "really pretty, but I don't want to go back." Considering the fact that more than a thousand people were vacated by army helicopter, that consensus will probably stand.

Persisting rains and fog grounded rescue craft for a period of time the following day, and search crews were sent out on foot to trudge along the muddy riverbanks in search of survivors. Four-wheel-drive vehicles were dispatched along the back roads to reach Drake, a hamlet halfway between Loveland and Estes Park. What they found was the remains of a wall of water which had run fifty feet higher than its usual tranquil flow, taking cars, trailer homes, livestock, and stores with it. They also found the gored remains of eighty people who did not escape the sudden onslaught of water.

Sometimes, the first few days after a flood are worse than the disaster itself. Coroners tried to match relatives' descriptions with police photographs. When they thought that they had matched a body with a description, they sent a clergyman to get a friend or relative involved, and only then were the next of kin allowed in to look at the deformed body. Dr. Charney, a specialist in disaster identification, related, "We're trying to minimize the emotional trauma. They're all battered. The action of the water was violent, flinging them all over the place. It happened very fast, and all we can do is guess. But it's our feeling that most people were knocked out first. It was quick and ferocious."

Survivors kept a vigil for several days after the incident at local morgues and hospitals. One man, Mike Watson, heard that his former wife, Cheryl, was going to a dance in the canyon. Only recently divorced, friends told him that they had turned around when the sudden downpour commenced but Cheryl and her companion had continued on. He called the hotel where they were to have stayed, and a clerk said she was safe. Later, he drove to the hotel to find that they had never been there. "I had always been just a daddy, now I've got to be a father." Divorced only three months before, he and Cheryl had become friends once more. "We were supposed to have a date this week." Cheryl Watson was never found.

It is sad that more lives were not spared and miraculous that many more were not taken. At the height of the tourist season, flash-flood warnings were issued well in advance of the torrential rains, which for the most part went unheeded. Some people waited three days for rescue, when they were airlifted under bright blue skies after many hours of chilly rain, rain which had swelled the twisting river out of its course Saturday night. Hundreds more were injured, several small communities were wiped out, and more than 100 million dollars of damage was incurred in this once-in-a-hundred freak of nature. At the ruined Big Thompson hydroelectric plant, beside the river fifteen miles west of Loveland, a Bureau of Reclamation sign stood as an ironic testimony to the tragedy. "Falling water is energy," it began.

The nation's attention was quickly diffused to Philadelphia, where the outbreak of Legionnaire's Disease had claimed at least twenty lives, and Ronald Reagan, in a last-ditch effort to secure the Republican presidential nomination, picked Pennsylvanian Richard Schweiker as his running mate. There was a severe earthquake in Peking, and the Viking satellite transmitted the first clear pictures of the Martian landscape. These were lively times indeed, but in Colorado, mountaineers were climbing all fifty-two of the state's "fourteeners" (peaks of 14,000-feet elevation) in search of victims rather than celebrating the one-hundredth anniversary of statehood.

26. FLASH FLOODS—WHAT TO DO

TERMS USED IN FORECASTS AND WARNINGS

Flash Flood means the occurrence of a dangerous rise in water level of a stream or over a land area in a few hours or less caused by heavy rain, ice-jam buildup, earthquake or dam failure.

Flash Flood Watch means that heavy rains occurring or expected to occur may soon cause flash flooding in certain areas and citizens should be alert to the possibility of a flood emergency which will require immediate action.

Flash Flood Warning means that flash flooding is occurring or imminent on certain streams or designated areas and immediate precautions should be taken by those threatened.

SAFETY RULES

Before The Flood know the elevation of your property in relation to nearby streams and other waterways. Investigate the flood history of your area and how man-made changes may affect future flooding. Make advance plans of what you will do and where you will go in a flash flood emergency.

When The Flash Flood Watch Is Issued listen to area radio and television stations for possible Flash Flood Warnings and reports of flooding in progress from the National Weather Service and public safety agencies. Be prepared to move out of danger at a moment's notice. If you are on the road, watch for flooding at highway dips, bridges, and low areas due to heavy rain not visible to you, but which may be indicated by thunder and lightning.

When a flash flood warning is issued for your area, act quickly to save yourself. You may have only seconds:

1. Get out of the areas subject to flooding. Avoid already flooded areas.
2. Do not attempt to cross a flowing stream on foot where water is above your knees.
3. If driving, know the depth of water in a dip before crossing. The road may not be intact under the water. If the vehicle stalls, abandon it immediately and seek higher ground—rapidly rising water may envelop the vehicle and its occupants and sweep them away.
4. Be especially cautious at night when it is harder to recognize flood dangers.
5. If you are out of immediate danger, tune in area radio or television stations for additional information as conditions change and new reports are received.

After the flash flood watch or warning is canceled stay tuned to radio or television for follow-up information. Flash flooding may have ended, but general flooding may come later in

A flood in the Black Hills of South Dakota resulted in this pile up on June 10, 1972. (*National Oceanic and Atmospheric Association*)

headwater streams and major rivers.

NOAA's Weather Service has helped set up flash flood warning systems in about 100 communities. In these, a volunteer network of rainfall and river observing stations is established in the area, and a local flood warning representative is appointed to collect reports from the network. The representative is authorized to issue flash flood warnings based on a series of graphs prepared by the Weather Service.

Courtesy of National Weather Service, National Oceanic and Atmospheric Administration

V. THUNDERSTORMS

Nothing in nature is so dazzling as a mid-summer thundershower, nor quite so loud and frightening. While thunderstorms can and do occur within the conterminious United States at any time of year, their favorite season extends throughout the summer months, with Florida ranking number one in frequency, more than eighty being reported there each year, while interior Alaska averages less than five. Most occur during the late afternoon and evening hours, as a result of convective heating, while showers associated with a frontal system can occur at any time, although they are most severe when enhanced by daytime heating.

One beneficial result of this natural phenomenon is the attendant rainfall. The height of the thunderstorm season coincides with the nadir of cyclonic activity across the United States, and many farmers, particularly in the Midwest and High Plains depend upon this blessing at the critical stage of crop development. Along the East Coast and throughout the Southland, atmospheric conditions favor the development of thundershowers, and the rainfall spans the months—which would otherwise be quite dry—between the end of the west to east cyclonic activity in the spring until the arrival of tropical disturbances in late summer and fall. Another tangible aspect of such activity is the fertilization of the earth, as lightning acts as an agent to break down the oxygen and nitrogen in the atmosphere to form nitrates which are introduced into the soil by rainfall runoff.

Thunderstorms form as a result of atmospheric instability, i.e., the continuous and steep decline in temperatures with altitude. Such conditions favor buoyancy in the atmosphere, so that a rising parcel of air is forced high enough to condense into visible water droplets (clouds), and the latent heat released in this process fuels the convection process and forces the clouds and water vapor to even higher levels, occasionally breaching the lower levels of the stratosphere, or 60,000 feet. Daytime heating is the usual spark for convective currents, although a hilly or mountainous range may also lift air currents orographically to such altitudes where the rising air is constantly surrounded by colder, denser air which feeds the convection process. Also, the convergence of airstreams in a frontal zone during a wintertime cyclone can create vertical currents which produce out-of-season thunderstorms.

Lightning is the exclusive by-product of a thunderstorm and is formed by the difference in electrical charges between the cumulonimbus cloud, which is positively charged, and the earth and upper atmosphere (particularly levels that are below the freezing mark), which have negative charges. Although the exact process by which lightning is formed is not known, it is believed that the splitting and collision of water droplets and ice particles, which concentrate in different parts of the cloud, form kind of a giant storage battery in the atmosphere. Only twenty percent of the available electrical charges reach the earth. In spite of this, it is estimated that, worldwide, lightning bolts strike the earth nearly 100 times per second. Over the past forty years, 8,500 people have been killed by lightning, while 25,000 have been injured.

Thunder can be heard as a distant rumbling, and is a harbinger of an imminent storm or can shake houses and rattle windows when directly overhead. It is caused by the rapid expansion of the atmosphere along the path of a lightning bolt and the incredible heat (of 10,000° C) generated by that impulse. Since the speed of lightning is so fast, it may be considered to reach the viewer's eye almost instantaneously, while the sound of the thunder plods along at a mere 1100 feet per second. The distance, in miles, of a lightning strike may be computed by dividing the number of seconds between the flash and thunder by five. Seldom is thunder heard for more than a distance of fifteen to twenty miles, and the average limit of audibility is about ten miles.

*Although individual storms have produced damaging hail and deadly tornadoes, most of the toll from thunderstorms results from forest fires, as a tree is lightning's favorite target. Each year, more than 7,500 forest fires are started by lightning in the western United States, as the height of the thunderstorm activity corresponds to their driest season. A thunder*storm*, incidentally, is distinguished by the National Weather Service from a thunder*shower *by the presence of rainfall in the latter case. Rainfall need not be present in a thunderstorm, as so often is the case in the West. Each year, nearly $25,000,000 damage is accrued from timber loss, while the twenty-two-year period between 1931 and 1953 resulted in the burning of more than two million acres of valuable timber—in square miles, an area larger than the state of Delaware.*

Over the years, there have been some flukish lightning events, many of them tragic. Perhaps the worst such tragedy occurred in Halifax Harbor, Nova Scotia, when lightning struck a French steamer loaded with 1,000 tons of munitions and exploded. More than 2,000 people lost their lives in the ensuing maelstrom, while another 3,000 were injured. The weather added to the horrors of the disaster that struck the Canadian province on December 6, 1917, for a raging blizzard was pounding the maritimes at the

"The Line Storm," by John Stewart Curry. (*New York Public Library*)

time. Over two square miles of land and building were laid to waste by fire.

Similarly, a single lightning bolt struck the former Naval Ammunition Depot, the nation's largest, at Lake Denmark, Morris County, New Jersey, in 1926 triggering explosions that were heard throughout five states and which caused $70,000,000 damage and took sixteen lives. A single bolt has been known to take the lives of more than 500 sheep and wipe out a flock of Canadian geese. Lightning was linked to the loss of the German zeppelin Hindenburg, *and modern airliners, although relatively safe, have not been totally immune to the destructive force of lightning.*

Particularly vulnerable to lightning are those involved in outdoor activities such as golfers (even Champion golfer Lee Trevino was struck during a tournament in 1975), baseball and football players, campers and hikers. The man who holds the world's record for enduring the most strikes is a park ranger by the name of Roy C. Sullivan, who was struck for the seventh time in June of 1977, earning him the title of "the human lightning rod." He has had his eyebrows seared and his hair set afire, suffered burns over most of his body, while only losing his big toenail in 1942.

The largest hailstone to fall in the United States measured 17.5 inches in circumference while weighing 1.67 pounds on September 3, 1970, near Coffeyville, Kansas. Similar sized stones *have been found throughout the High Plains, causing an average of $773 million dollars' crop damage throughout the United States every year. Perhaps one of the worst storms in history occurred at Seldon in northwest Kansas on June 3, 1959, when depths reached an astounding eighteen inches of mid-summer ice on the ground. Throughout the Midwest, accumulations of hail have been known to drift and remain on the ground (in protected areas) for more than seven weeks, while snowplows are occasionally brought from their winter mothballs to clear the ice-clogged roads after a storm.*

While dangerous and potentially destructive, thunderstorms are a common phenomenon throughout the summer months and provide one of our most dazzling glimpses of Nature. Many principles concerning the dynamics of meteorology are present in the common thunderstorm, but its most immediate and welcome relief is the sudden drop in temperature that accompanies the storm's cold downdrafts. It's the next best thing to an ice cream cone.

27.
FIREWORKS AT LAKE DENMARK

In the course of a day, throughout the world, lightning strikes the earth tens of thousands of times. Most of them are beneficial, and it is rare when a particular stroke causes as much havoc as the town of Lake Denmark witnessed on the tenth of July, 1926. This was the day of the unimaginable, when nature declared war on a munitions plant in northern New Jersey and America's fighting men laid down their arms in terror. Coming only six days after Independence Day, it was a belated fireworks display like none other ever experienced in America.

An unusually strong cold front accompanied by sixty-mile-an-hour gales and rapidly falling temperatures blasted through the New York metropolitan area in the late afternoon, uprooting trees and unshingling roofs as the hottest day in fifteen years was coming to an end. In the town of Lake Denmark, at the present site of the Picatinny Army Depot, personnel were settling down to dinner as dark clouds loomed on the western horizon. For years this site had served as the nation's largest munitions plant, and safeguards had been taken to avoid the possibility of fire, for accidents were not within the realm of reason. One hundred eighty-four buildings stood in the newly expanded facilities on several hundred acres of land, and it was supposed that they stood far enough apart to defend against a major catastrophe should the unthinkable happen. No one ever conceived of the disaster that was to descend from the skies on this stormy evening.

Contained within these buildings were 90,000 pounds of black powder and an additional 50,000 pounds of TNT. Suddenly, from directly overhead, there was a violent clap of thunder and a vivid flash of lightning. A red ball of fire accompanied by a booming noise that shook the countryside for thirty miles around was seen leaping towards the sky. A major storage building had been hit, setting off a series of fires and explosions that would not diminish for three and a half days. A pair of naval officers who had been on afternoon leave and were now entering the depot described the scene: "The first explosion shattered the windshield of our car, and before we had come to our senses a second explosion overturned the car and we crawled out, all cut up. A third explosion sent us rolling into a ditch, and we crawled the rest of the way to Picatinny on our hands and knees, with stuff shooting over our heads while the heat was becoming unbearable."

At the first sign of smoke, fire alarms throughout the depot were set off, and men came running to the scene of the disaster. The scene, in one man's words, "was like a touch of hell" with artillery shells whizzing around at random, some even flying over the nearby hills and shattering every window within a ten-mile radius. Captain O.C. Dowling, who was one of the first on the scene, lamented, "There was nothing we could do to fight the fire, we could only run before it." One man, who was only identified as Brundage was on kitchen duty when the blast erupted and was literally blown out of his boots, fifteen feet into the air. One of his companions related, "Brundage and I and this other fellow headed across the hills and were nearly three miles away before Brundage discovered that he had lost his boots in the explosion and was traveling in bare feet." Another opined, "There'll be a lot of them wandering around through the mountains tonight, dazed from concussion, and they might not get back until daylight. The woods are dense here, and if a man doesn't know them he'll never find his way out."

The calamity was not confined to the depot. Villages ten and fifteen miles away were shelled for three days, with some villages totally ruined, windows broken thirty miles away, and cars hurled off the highway and houses torn from their foundations. Each time the booming seemed to subside, the ammunition heated up to the point of explosion and a fresh bombardment ensued, with vast sheets of fire lighting up the sky for miles around. Bathers reported that a lake sixteen miles from Picatinny rose four feet the instant of the first explosion, and swimmers were covered with a rain of black powder and splinters of wood. Some parts of the depot buildings were hurled as far away as Oakland, thirty-two miles from Lake Denmark, as though by the power and whimsy of a tornado. Some of the shells screamed like sirens, resembling a most spectacular fireworks display.

Nearby hospitals were alerted, and every available doctor and nurse from as far away as Brooklyn were called into action to care for the wounded. One man, dazed by the shock of the blast, entered Morristown Hospital and sought to identify his infant son as "Baby." "Whose baby is it?" asked a sympathetic nurse.

"Mine," said the man.

"What's your name?"

The man stared blankly. "I don't know," he said.

"What's the baby's name?"

"I don't know," the man replied again.

The nurse turned away for a moment and he disappeared.

When it finally ended, four square miles in and surrounding Picatinny resembled a large, lifeless crater of the moon. More than $93,000,000 damage was assessed, and sixteen people died as a result of a single lightning stroke. Most had to flee to the hills for cover, while some swam across Lake Denmark to escape danger. For those who had fought in World War I, the scene resembled an artillery-swept field in France. One man, who had been eating at the time the bolt flew out of the sky, came back four days later to assess the damage. Several hundred yards from the dining hall, he found some belongings of his and a couple of pork chops that were to have been his dinner four nights earlier. He picked up the chops and started nibbling on them, shaking his head.

28. THE WRECK OF THE ZEPPELIN *HINDENBURG*

EVER SINCE MAN HAS DREAMED OF FLYING, THE DANGERS OF AVIATION HAVE BEEN APPARENT. The Wright brothers experienced their share of failure before their successful flights from Kitty Hawk in the opening years of the twentieth century. The concept of flying by means of a lighter-than-air craft had been popularized in the middle of the 1800s by Count Zeppelin of Germany in his aviational experiments. The first three decades of the twentieth century witnessed many flights, some of a commercial nature, on a zeppelin-like vehicle, predating modern aircraft by a generation.

Considering the number of fatalities resulting from these flights, it is surprising that the technology was still alive and well at the demise of the *Hindenburg* in 1937. The list of catastrophes was impressive. On July 15, 1919, a British NS-11 fell from the sky during a violent thunderstorm into the North Sea. Six years later, on September twelfth, a seventy-mile-an-hour gale ripped the American-made *Shennandoah* into shreds, dashing the ship 7,000 feet to the ground, killing fourteen, while twenty-seven miraculously escaped. Later in that decade, on May 25, 1928, the zeppelin *Italia*, that was on an expedition to the North Pole, blew up in the Arctic, killing most aboard, yet some of the crew managed to stay alive for three months before being rescued in a latter-day *Alive* adventure. In 1933, a vicious thunderstorm wrecked the U.S.S. *Akron* off Barnegat Lightship, killing all of the crew's seventy-three passengers. Only two years before the wreck of the *Hindenburg*, the *Akron*'s sister ship the *Macon* had plunged into the Pacific during a storm, but all were saved. The following year, in 1936, the *Hindenburg* had its maiden voyage across the Atlantic and was to make another nine round trips across the ocean in an average time of sixty-one hours. The captain of those flights was aboard the *Hindenburg* in an advisory capacity when the ship met its doom on May 6, 1937. His dying words were, "I couldn't understand it." Captain Ernst Lehmann was a respected flyer of his day, adulated among the public as a true pioneer.

A protracted heat wave had been plaguing the eastern third of the nation on that sticky afternoon. A sharp cold front had just pushed a line of thunderstorms through the area in the early evening hours, delaying the landing of the *Hindenburg* at a naval air station in Lakehurst, New Jersey. Earlier in the day, the craft had passed over Portland, Maine, Boston, New London, and New York City at a relatively low altitude to throngs of enthusiastic crowds. The game between the Dodgers and the

The flaming *Hindenburg* floats gently to her death. (*Department of the Navy*)

Pittsburgh Pirates was momentarily disrupted as the *Hindenburg* made a direct pass over Ebbetts Field in Brooklyn. Tens of thousands of people witnessed this most famous air ship's last journey.

When it appeared from a fog bank over the naval air station its altitude was much too high to attempt a landing, and it circled around for another pass. Its second approach was marred by its forward speed which, according to eyewitnesses, was two or three times its approach speed. Nevertheless, it dropped its two lines and began venting a gaseous hydrogen substance for its final descent. Both lines were caught and moored, but the captain, alarmed by the forward drift, shouted, "More pay out! More pay out!" One of the landing crew heard him and released the line, but the other did not. Consequently the *Hindenburg* lost its poise and turned its nose section upwards, while its rudder scraped along the ground. No one knows exactly what happened next.

Some say that in the pilot's urgency to slow the craft down he revved the ship into reverse, causing a backfire which threw the deadly spark. Some believed that residual electrical charges or a distant lightning bolt ignited the deadly hydrogen protons, and some say it was an act of sabotage. Inquiries after the accident failed to resolve this issue, but almost all who witnessed this event, including a ground crew of nearly 200—as well as those who were there for the return flight, agreed that there was a violent explosion and spectacular fire occurring right over their heads. One man screamed, "Run for your lives," and the packed audience beat a retreat while the zeppelin flew apart, as though made of paper-mâché.

The touch down in a fiery blaze. (*Department of the Navy*)

The flash set off a thrill of terror among spectators as the rear gondola burst into flames and quickly enveloped the entire craft in a blazing inferno. There was something slow and graceful about the craft's demise; many did not realize for the moment that they were witnessing a tragedy taking place before their very eyes. It came down deliberately and settled so quietly that one survivor related, "Even in breaking up, the *Hindenburg* was gentle to its passengers—those who escaped alive."

Another man, a German circus acrobat, told his shocking story:

> I was just packing my belongings and felt a slight tremor shaking the ship. There was an explosion that sent the tail of the airship to the ground...the effect from where I stood was like an air pocket. As I walked toward the promenade to check, a second tremor occurred.
>
> There was very little confusion among the passengers, no screaming and hardly any noise. Nobody knew what had happened...they were just curious. When I reached the promenade deck, the ship was 20 feet off the ground. Although I wasn't alarmed, I knew something was wrong, and I decided to get off as fast as I could. I climbed to the sill of the window planning to jump when suddenly the steel under my feet buckled and catapulted me to the ground.

Later, a reporter asked him, "How on earth did you do it?" Feeling himself to see if he was all there, he responded, "I don't know. Whew! Am I in luck—not a scratch." Only minutes before, he had

EXTRACTS FROM LOG OF THE HINDENBURG
FRANKFURT, GERMANY ~ TO ~ LAKEHURST, N.J
17~20 MAY, 1936

PLATE ~ I

May	Time GMT	Time Local	Alt. Metres	Temp. (C.) Air	Temp. (C.) Gas	R.H. (%)	Position Lat.	Position Long.	Course	Drift	Str Course	Wind Dir.	Wind Force (mps)	Grd Spd.	Time (Min)	Dist.	Total Distance	Remarks
17	[illegible]	[illegible]	-	11	15.4	81	Hangar - Off Mast		-	-	-	N	3-4	1				Barometer 752.3 (mm) - Mostly clear sky - Patches ground fog
	[illegible]	[illegible]	-	12	14.7	[illegible]	" - Walked Out		-	-	-	N	2-1					
	[illegible]	[illegible]	-	11	[illegible]	[illegible]	Took Off		-	-	-							
	[illegible]	0644	[illegible]	13	12.0	[illegible]	Tilburg, Germany		-	-	-	SSE	5			164		
	0830	0830	[illegible]	-	-	-	Dover Cliffs		-	-	-	-	-	-	-	-	228 mi. (at 0730)	Scattered High Clouds. - Standing down English Channel, haze and fog.
	[illegible]	1000	200	16	18	56	50° 33'N	0° 25'W	280	-	-	-	-	76	70	65		Fog - Extending above ship
	[illegible]	1100	200	11.8	18.9	65	50° 13'	1° 52'	"	-	-	NW	2-3	60	60	60	508	4/10 ACu. to West - Hazy (Heavy).
	[illegible]	1200	150	11.0	16.6	70	49° 51'	3° 24'	"	-	-	NW	2-3	63	"	63		[illegible]/10 ACu. to West - Hazy.
	[illegible]	1400	150	11.0	14.2	66	49°09'	6° 14'	"	-	-	NNW	3-6	60	"	60		7/10 Ci.St and ACu.
	1600	1600	150	10.0	13.0	66	48°28'	9°07'	270	-	-	WSW	3	60	"	60	750	8/10 ACu. - Good Visibility
	1900	1800	220	9.0	11.5	[illegible]	48°40'	13°52'	"	-	-	NNE	18-20	75	"	75		5/10 Cu + CuNb. - Showers in vicinity - Passed through Occluded Front at 1700
	2100	2000	220	9.0	10.8	84	48°07'	16°38'	241	-	-	NNW	18	65	"	65	1071	10/10 CuNb - Rain
	2300	2200	250	9.0	10.8	80	47°00'	20°08'	"	-	-	N	12-13	72	"	72		10/10 St and StCu.
18	0100	0000	250	7.5	9.5	82	45°18'	24°29'	"	-	-	N	6-7	[illegible]	"	70	1352	8/10 Cu and St.
	0300	0200	250	9.0	10.0	82	44°05'	27°30'	"	-	-	NE	3-4	[illegible]	"	72		10/10 St. - In Clouds
	0500	0400	250	12.0	11.0	[illegible]	44°00'	30°30'	270	-	-	ESE	2-3	[illegible]	"	67	1703	8/10 St. - Rain
	0800	0600	600	12.0	11.0	80	44°00'	33°28'	"	-	-	SE	7-8	[illegible]	"	66		[illegible]/10 ACu - Fair Visibility.
	1000	0800	800	13.0	16.3	88	43°47'	36°10'	240	-	-	SW	15	[illegible]	"	[illegible]	1931	[illegible]/10 ACu. - [illegible]/10 CuNb.
	1200	1000	130	18.0	18.0	83	[illegible]	[illegible]	"	-	-	SSE	15-16	[illegible]	"	[illegible]		[illegible]/10 StCu, ASt + ACu - Approaching Front.
	1400	1200	130	18.3	17.1	82	[illegible]	39°03'	"	-	-	S	21	[illegible]	"	[illegible]		[illegible]/10 Cu, St + StCu - CuNb ahead - Heavy haze.
	1600	1400	[illegible]	17.0	18.6	84	[illegible]	39°00'	-	-	-	SSW	19-20	[illegible]	[illegible]	[illegible]		[illegible]/10 Cu + CuNb - Rain - Flying South looking for hole to get thru Front.
	1800	1600	[illegible]	18.3	16.3	[illegible]	41°16'	[illegible]	249	[illegible]	276	WNW	6-7	[illegible]	[illegible]	[illegible]	2239	[illegible]/10 Cu + ACu - Wind decreasing - Passed thru Front.
	2000	1800	300	14.0	15.0	64	[illegible]	[illegible]	"	6	263	SSW	6-7	[illegible]	[illegible]	[illegible]		[illegible]/10 Cu - Visibility improving.
	2200	2000	[illegible]	13.9	14.9	62	[illegible]	[illegible]	"	[illegible]	[illegible]	SW	[illegible]	[illegible]	"	30		[illegible]/10 St. - Ship in clouds.
	0000	2200	[illegible]	16.0	[illegible]	[illegible]	[illegible]	[illegible]	"	30	[illegible]	SSW	17-18	[illegible]	"	42	[illegible]	10/10 Overcast - Clouds under ship.
19	0300	0000	[illegible]	13.8	16.0	[illegible]	[illegible]	49°13'	285	3	288	SW	[illegible]	[illegible]	"	30		T/S to StBd - Following along Front, seeking favorable place to go thru
	[illegible]	[illegible]	[illegible]	18.0	17.4	[illegible]	40°10'	[illegible]	270	[illegible]	250	WSW	18-19	[illegible]	"	28		[illegible]/10 Ci St.
	0700	0400	[illegible]	18.3	18.4	84	40°02'	[illegible]	233	1	234	SW	18	28	"	28	2764	[illegible]/10 St.
	0900	0600	200	19.0	19.0	93	39°27'	[illegible]	"	1	236	SW	10	34	"	34		[illegible]/10 St. (Fog).
	1100	0800	200	20.0	20.5	68	38°43'	55°30'	"	3	[illegible]	WSW	16	[illegible]	"	35	2898	[illegible]/10 ACu + CuNb - Fog under ship.
	1300	1000	200	19.0	19.5	85	38°00'	55°15'	270	10	260	WSW	12	40	"	40		[illegible]/10 St + StCu - Heavy haze.
	1500	1200	1280	16.5	15.5	52	38°00'	[illegible]	"	3	273	W	6-7	49	"	49	3069	[illegible]/10 CiSt. - Ship flying in broken St.
	1800	1400	1400	15.8	17.1	41	38°00'	[illegible]	"	4	266	WSW	7-8	53	"	53		3/10 Ci, ACu. - Lower Cu.
	2000	1600	1400	16.3	17.3	43	38°11'	[illegible]	282	15	267	SW	10-11	52	"	52	3330	2/10 CiSt - 3/10 Cu.
	2200	1800	200	12.5	23.5	42	39°11'	66°30'	277	22	255	SW	[illegible]	55	"	55		7/10 ASt + St.
	0000	2000	200	20.5	22.5	66	39°24'	66°50'	277	31	246	SW	18	60	"	60	3541	9/10 ASt.
	0200	2200	200	18.3	20.0	75	39°34'	[illegible]	277	32	[illegible]	SW	19	40	"	40		10/10 Upper + Lower Clouds. - Hazy.
20	0400	0000	300	20.0	[illegible]	80	39°36'	[illegible]	277	20	257	WSW	19	30	"	30	3690	Cloudy - Hazy.
	0600	0200	300	19.0	18.5	62	39°34'	[illegible]	280	5	275	W	15	35	"	35		7/10 Ci.
	0800	0400	300	18.0	17.0	50	39°35'	[illegible]	273	15	288	WNW	19	30	"	30		Few Ci.
	1100	0600	Landed at Lakehurst, N.J.															Few Ci.

Log extracts from the *Hindenburg*. (*Department of the Navy*)

been one of the ninety-eight people aboard seen laughing and waving their hands out the windows to the crowd below.

Sixty-two people escaped with their lives either by jumping or by being blown out of the ship by the blast of the explosion, while thirty-six died in the smoldering ruins of the *Hindenburg*. The raindrops which still fell at the time of the disaster in all likelihood coated the craft like a giant envelope of static electricity, while contact with the wet rope which was to serve as a landing line probably caused the lethal spark, with the man on the ground serving as a negative charge, in this instance. Nevertheless, millions of Americans flocked to their local movie theaters the following day to witness for themselves this most widely seen incident. President Roosevelt expressed his condolences to the premier of Germany, one Adolph Hitler, and to the families of those deceased, mostly Germans. Hitler and his peer among peers, General Hermann Goering, thanked the American people for their daring in carrying out the rescue effort. In a matter of only a few years, these men would not be so grateful nor sympathetic, but in spite of man's accumulated knowledge and cunning devices, Nature had the last word, for the inevitable could not be curbed or forestalled.

29.
THE TRIPLE TRAGEDY OF '63

PERHAPS THE MOST COMMON MEMORY OF OUR LIVES WERE THE HOURS AND DAYS FOLLOWING the assassination of President John F. Kennedy. Nearly every human heart was reduced to sorrow under leaden November skies after the sudden and untimely end our president met. Late November is the year's own time to sleep, a farewell to the lush green pastures of summer and the crisp fiery autumn vegetation by the roadside. While the bare limbs and branches of November bent and braced themselves for the buffetings of winter, the mournful wind teeming through the woods seemed sensitive to human emotion. In the closing days of 1963, we lived through an eternity, though we found ourselves mortal. We persisted, as Hal Boland of New York *Times* wrote, "...with a capacity to dream and hope and dream again. Man, with a capacity for shock and grief, but also with his inheritance of faith, of belief, and participation in the great truth of continuity." The nation's shock and disbelief was not alleviated in the ensuing days, for a series of tragedies were about to strike and set the tone for the balance of this tumultuous decade. But the sun did not stray from its course, and the stars continued to shine, and the rugged hills did not perish. Take away everything else, and the human eternities remained."

Thanksgiving Day was marked by a pall of sorrow, and on the following day, November twenty-ninth, news from Canada flooded the papers regarding a jetliner that had crashed following take-off in a drenching rainstorm just north of Montreal. It was the worst air tragedy in that country in history—claiming all 118 lives on board—and exceeded only by the disaster at Olny Airport in Paris in June of 1962 with respect to casualties. The craft had left Montreal for Toronto only four minutes before its demise when its engines conked out, either as a result of the heavy rains or after being struck by lightning. Since there was no communication with the plane following take-off, one can only surmise by the eyewitness accounts from the ground:

"I heard a terrible explosion and saw a huge red ball of fire in the air. The plane crashed just about right away after that and my husband called the police." Others told of hearing an explosion and seeing "a long red streak in the sky." The red-trimmed, silver jet smashed into the countryside near Ste. Therese de Blaineville, between Montreal and the Laurentian Mountains, digging a huge crater into the ground which soon filled with rainwater. The following day, the storm system that had been centered over Pennsylvania and had generated the Canadian rains, pulled down enough cold air to turn the precipitation to snow throughout the Northeast, with an inch accumulation reported in the New York City area.

Only a week later, catastrophe was to surface again, this time in the rolling hills of northeastern Maryland, near the town of Elkton. A huge Pan Am jet bound for Philadelphia from San Juan had just taken off from Baltimore after seventy-one people deplaned in that city. The craft was confronted with a violent thunderstorm with fifty-mile-an-hour headwinds at about the time lightning flashed out of the sky, about ten miles southwest of Wilmington, Delaware. The last words radioed from the doomed craft were, "Going down in flames." Witnesses reported seeing a bright flash lighting up the sky "like dawn" and heard two explosions. The fragments of the plane fell over a two-mile-wide area between Route 40 and the Northeast Expressway, on what would have been a twenty-minute flight to Philadelphia. The Gregg family of Elkton had been watching television when the explosion hit. Having seen the burning wreckage in the sky, they immediately rushed to the basement just before the plane landed in their backyard, only fifty feet from the house. Bits of debris showered down upon the house in the moments before the crash. "At first we thought it was an atomic bomb. Then you could actually see people falling out...The plane came down slowly, and when it hit the ground, it exploded again."

This incident ended a perfect safety record for the carrier that had logged over 712 million passenger miles and flown eleven million passengers safely over the years. It is one of the few major airline disasters attributed soley to lightning. So often major weather events are linked to airline disasters that the list seems endless. Only four years later, in January of 1967, three U.S. astronauts—Grissom, Chaffee, and White—were to die in their *Apollo I* spacecraft (on the ground) while Chicago's deepest snowfall was in progress. (See section on Snowstorms and Blizzards.) As recently as January of 1982, seventy-nine lost their lives when a jetliner out of Washington, D.C., crashed shortly after take-off in a blinding snowstorm. In the summer of 1982, 130 more died when their plane went down in a thunderstorm near New Orleans.

The winter of 1963-64 proved particularly harsh, as major snowstorms routed the East Coast on January nineteenth and February third, depositing a combined total of over three feet. The great northeast drought of the early and middle 1960s was already well underway, and reservoirs had dipped to between thirty and thirty-five percent capacity. This two-week period in late autumn of 1963 proved remarkably memorable, as the forces of Nature ganged up relentlessly upon a people with grief enough to bear.

30.
THE SECOND TIME THE LIGHTS WENT OUT

ONE OF THE IRONIC FEATURES OF OUR TECHNOLOGICAL AGE IS THE POSSIBILITY FOR THE entire system to go wrong, leaving the sorcerors baffled and unable to immediately correct the situation. One such extreme example occurred on the evening of November 9, 1965, when nine northeastern states, eastern Canada, and parts of the Midwest were totally blacked out by a mysterious power failure that lasted approximately fourteen hours. Thirty million people were literally stopped in their tracks, including an estimated 80,000 in elevators, subways or trains. It took days to trace the source of the problem, and despite the uncertainty, people took the event with good naturedness and a decided lack of the crime and looting that so often accompany an event like this. In fact, hospitals reported record numbers of childbirths the following summer, nine months after the lights went out.

Twelve years later, in July of 1977, the lights again flickered and vanished over Gotham, but the results were entirely different this time. Mayor Beame of New York dubbed it "a night to remember," as a series of lightning bolts triggered the most massive looting in the nation's history, an estimated billion-dollar loss, in the overnight darkness of July thirteenth. A severe thunderstorm was in progress, spawned by an intense low-pressure system over the Great Lake region. Lightning struck from the thunderclouds over the Indian Point Nuclear Power Plant, owned by Con Edison, in northern Westchester County. A company spokesman explained that the major transmission lines were repeatedly struck, causing their failure, and the shortcircuits that resulted trapped the relays and ultimately shut down transformers and power stations throughout New York City.

The sky lit up in the vicinity of Buchan, near the Indian Point plant, at 10:45 P.M. Residents reported seeing a strange light and hearing a "whirling sound" that lasted for about ten seconds. The power immediately faltered in Westchester and Long Island, and Con Edison workers went into action, trying to save the power for nine million New Yorkers by a series of temporary brownouts and blackouts in the suburban areas. One by one, however, major power stations failed as lightning bolts continued to hit their 345,000-volt power lines. Within an hour of the first failure, "Big Alice"—a massive generator in Dutchess County—shortcircuited, and within minutes lights all over the city darkened and plunged New York into mayhem. Within minutes, all three metropolitan airports were closed for business, and thousands of bewildered New Yorkers became

Lightning strikes the tower of the Empire State Building. (*New York Public Library*)

trapped in subways, elevators, and trains. An airline pilot, circling over Queens, inquired, "Where is Kennedy Airport?" as the landing strip disappeared beneath him. The tower told him to go on to Philadelphia. The pilot persisted, stating that he had a cargo of perishable strawberries that were to be delivered to the New York markets. "What am I supposed to do with the berries," he asked plaintively. "Eat them," came the reply.

Another man, dressed in a business suit, was seen to exit a stalled subway in Harlem at about 9:20 P.M. With him he carried a brand new set of golf clubs and, anxious to try them out, he groped across the darkened tracks and struggled up to the middle platform. Once he had his footing, he placed his feet squarely on the cement strip on the other side and proceeded to take his practice swings. Liv Ullman, starring in Eugene O'Neill's *Anna Christie*, finished her performance in candlelight, when the blackout descended, and the audience rose to a thundering ovation at the final curtain. The show must go on.

Beneath the street level, subway cars came to an eerie halt, only moments after they had been alerted of the impending crisis. One of the stranded victims related, "They told everyone from the back cars to move to the front and we climbed along the catwalks holding onto the railings. It took about 45 minutes until we got out. I think most people seemed to be enjoying themselves." Back on the street, in the pitch blackness, a woman was walking her dog when a man whistled at her and yelled,

"Hey, Beautiful!" Her deft reply was, "How can you tell."

Hospitals had to resort to emergency-backup power generators, and when these failed, emergency operations were conducted outdoors under flashlight. Despite the successes, two patients were lost with this procedure. Doctors and nurses had to resort to artificial breathing devices, powered by human hands, when respirators failed. These had been ordered after the 1965 blackout and yet, despite that investment, doctors and nurses were seen stumbling around in the darkness, working with flashlights and lanterns, bumping into beds and into intervenous equipment. It's astounding that the loss of life in the cities hospitals was not any larger.

The Statue of Liberty was about the only island of serenity on the long summer night of July 13, for its automatic power was the only bright spot in the otherwise total blackness of New York City. Within a few minutes of the power failure, glass shattered and sirens wailed, setting the scene that was to produce the largest police roundup and greatest loss from looting in one night in the history of the country. Nine-hundred fires were set, and about three times the usual number of alarms were pulled on that ruthless night. Firefighters contended with snipers, bottles, and bricks. More than a hundred police and firemen were injured, while a total of 3,776 arrests were made. Nearly every cop from a force of 25,000 was on duty that night, and they were powerless to stem the wave of crime that was sweeping many sections of the city. Particularly hard hit were businesses in Harlem, East Harlem, the Bronx, Bedford-Stuyvesant, Jamaica, and Flatbush. Hundreds of miles of city streets were littered with glass and debris, as the plunderers favored television sets, food items, and furniture.

An auto dealer in the South Bronx reported that his steel doors and steel-edifaced windows had all been smashed, and that fifty brand-new cars had been driven away. The arraignment process had been snagged in the slowness of checking out police records brought about by the failure of the computer in Albany to respond...as a result of the earlier lightning strikes that had caused the blackout. Police radios crackled with additional reports of crime throughout the night and well into the next day. Residents grew accustomed to the metallic clang of trash cans used to demolish store keepers' metal shutters. The vandals, emboldened by the darkness, focused their attention on the smaller shops and managed to get away with more than a billion dollars' worth of goods.

Nineteen-hundred-seventy-seven had been described as producing some of the most bizarre weather of the century, the world over. The words *coldest*, *hottest*, *driest*, and *wettest* were sprinkled practically daily in the world weather reports. Crop failures throughout the world were becoming rampant, and temperatures in the high nineties accompanied by high humidities were attributed to the breakdown of law and order on that mid-summer night. Unlike the cool climate that prevailed in November of 1965, power failures sent people streaming into the streets the night that the glittering skyline of Manhattan vanished. Come the morning light, the looters scattered like cockroaches after looting nearly 2,000 stores on one busy night. While the terror of the night before had diminished, the anguish lived on, as residents had no water, no air conditioning or vital services; taxis gouged passengers, while Wall Street suffered an estimated twenty-five million dollars lost in commissions. All of this as a result of a thunderstorm, and the blighted story of a lone engineer who struggled and failed to save New York City's power. His supervisor remarked, "The fellow feels terrible."

VI. HEAT WAVES

A period of unseasonably hot and frequently humid weather will strike some portion of the continental United States virtually every summer—a heat wave—and in a "normal" year, at least 175 Americans will succumb to the rigors of heat stress. Among the large family of natural hazards—hurricanes, tornadoes, lightning, floods, and earthquakes—only the cold of winter is responsible for more deaths. In the forty-year period from 1936 to 1975, more than 20,000 people in the United States were killed due to the combined effects of solar radiation and heat. The ugly summer of 1936 when 4,678 lives were claimed by the heat, and of 1952 when more than 1,400 people died, as well as ten other recent summers when the numbers were far greater than normal made the averages climb to over 500, nearly triple the expected rate.

As recently as 1980, heat and drought contributed to 1,265 lives and cost the nation nearly twenty billion dollars, according to the Environmental Data and Information Service of N.O.A.A. Hundreds of miles of major highways buckled, and water resources in many parts of the country were seriously depleted. Most of those who died were either elderly or poor and lived in central-city apartments, deprived of air conditioning and ventilation. The highest number of fatalities occurred in Missouri, although many other states had higher temperatures for longer periods. The heat wave began in mid-June as temperatures exceeded 100 degrees every day for nearly two consecutive months in the Southwest states of Texas and Arkansas. By mid-July, the heat wave spread eastward to the Ohio Valley and Mid-Atlantic states, and remained entrenched with only minor respites through the first week in September. Many midwestern and eastern cities recorded all-time-high readings, and the nation's capital on September fourteenth recorded its sixty-third day of the season with readings over ninety degrees, while Little Rock had its forty-third day with temperatures over a hundred.

The important factors contributing to heat stresses are: 1) air temperature, 2) humidity, 3) air movement, 4) radiant heat from incoming solar radiation, 5) atmospheric pressure, 6) physiological factors which vary among the populace, 7) physical activity, and 8) clothing. Temperature and humidity are the two key fac-

An Oklahoma farmer struggles to salvage his automobile after a severe dust storm in 1934. (*National Archives*)

tors, and they can be controlled by air-conditioning. It is in the absence of air conditioning that the other factors come into play. To relieve conditions in the summertime, particularly after the second or third day of a protracted heat wave, physical activity must be reduced, one should avoid prolonged outings in the direct sunlight and wear lighter clothing. Just as water evaporates from your skin and causes a distinct chill on your skin when you step out of a shower in winter, temperatures above body heat heat your system up, in varying degrees, depending upon relative humidity. The apparent temperature felt by your body can rise as much as fifteen degrees from the actual temperature, depending upon humidity and sunshine levels. Humidities of fifty and sixty percent combined with temperatures over ninety inhibit your body's cooling system (i.e., sweating) from working effectively and put individuals in the heat-stress danger zone.

There is nothing in American weather annals to better match the heat waves which touched off the dust bowl droughts of the 1930s than those which struck the country's Plain states in 1930, 1934, and 1936. Each of these years brought progressively more severe summer weather. Those years alone were responsible for at least 15,000 deaths, owing to the cruel conditions which prevailed. All-time temperature records were established in those years in North Dakota, Kansas, South Dakota, Oklahoma, Arkansas, and Texas when temperatures soared above 120 degrees. In the Midwest and east, temperatures above 110 degrees saw records tied or broken in Indiana, Louisiana, Maryland, Michigan, New Jersey, Pennsylvania, West Virginia, and Wisconsin.

The culprit is almost always a stagnant air mass centered across the nation's Southwest region which brings the famous red Santa Ana winds of the Far West or the equally prominent Bermuda High, which can prolong hot, miserable weather in the East for more than a week or two. One need not look further than the hot summer of 1955, which ended in the twin hurricane phenomena of Connie and Diane. The Santa Ana winds of the West generally come in September, bringing with them air nearly too hot and dry to breathe, like those which brought very hot summers from 1955 through 1961. The Santa Ana winds which blew through the mountain passes in November of 1969 brought the lowest average humidity in nearly twenty years, and the desert-like air blew 100 degrees-plus temperatures on the wings of seventy-eight mile-an-hour winds.

With the advent of air-conditioning, many heat-related deaths of the past need not be repeated, if everyone had access to this during the critical days of a heat wave. Unfortunately, people still do not understand the limits of the human body in times of heat stress, and the tragedies of the past will be repeated. Even the statistics reported here are probably much too low, as many deaths ascribed to cancer or heart failure, stroke, and influenza were heat-related, as these statistics tend to rise threefold in a prolonged wave of hot weather.

31.
HOT AS THE FOURTH OF JULY

In the years preceding the World War I, America paid little attention to the occasionally disturbing news from the European continent but instead relished the feats of their own national heroes. Ty Cobb was on top of the list with his hefty .430 batting average, and the passing of one of the lustiest nineteenth-century gamblers, a man by the name of Davy Johnson, merited more attention than the skirmishes and preparedness for war on the European stage. Johnson was the last of a breed which flourished in America during the 1890s when pool halls became gambling casinos, featuring roulette, poker, and "turf" or the horse races. Johnson was known for his extravagance, his lavish tips and gifts, and for the ladies who always accompanied him. The "plungers," as they were known, made a daily living on wagers, and it was always said that his best friend was his horse. No one ever bridled at his behavior.

Flying captured the public's imagination at the turn of the century, and a decade after the maiden flights by the Wright brothers, in 1901, big prize money was at stake, as many intrepid pilots flew prototypes of the *Kitty Hawk* for hundreds and, occasionally, thousands of miles. In the summer of 1911, the New York *Times* offered a cup for the first successful flight from Boston to Washington, D.C., and Harry Atwood was among the favorites, having accomplished the first leg of the journey, a distance of 93.2 miles at an average speed of seventy-eight miles an hour. Much like the late Davy Johnson, the notions of pleasure, excitement, and speculation made flying glamorous, and papers throughout the country followed this race on their front pages. For many, the sight of these webbed planes swooping over their towns and cities was a first, and brought them perhaps as close to a plane as they would ever be.

The middle of the country was heating up, late in June of that year, and the sizzling heat progressed eastward across the Central Plains, through the Middle Mississippi Valley, and on to the Ohio Valley. By the first of July, 1911, practically the entire nation east of the Rocky Mountains was under its spell as temperatures soared well into the nineties. The second of the month was rated as the hottest day in nearly twelve years in many eastern cities, and the humidity began to climb. "It's hot," volunteered the weatherman in New York, but confiding to a reporter that he felt sorry for his colleague. When the newsman asked why, he was told that his colleague had the bad sense to vacation in Florida, where temperatures were over 100 degrees.

Amidst the stifling weather, a breeze clocked at three miles an hour was about as effective at cooling as a woman's lace fan.

Some of the first creatures to suffer were the polar bears in the zoo. While the rest of the animals were being sprayed with cool water, the polar bears vied with each other for possession of the cage's only cold-water tank. There was considerable grumbling and growling from those quarters. Equally entrancing was a young woman who strolled into Coney Island Park wearing purple pantaloons and a harem skirt. A considerable crowd assembled behind her, shouting, "Suffragette! Dress Reformer! Trousers for Women!" Unabashed, she strolled over to the dance pavilion with her companion and began to dance. By now, the sight caused people from the carousel and the scenic railway to throng over to digest the scene, and concessionaires grumbled to policemen that the young woman was stealing their business. The police asked her to leave. Sixty years later, a buxom miss was to capture Wall Street's attention each day as she strolled out of the subway on her way to work before ever-increasing thousands of leering men.

Meanwhile, new temperature records were falling in the eastern half of the nation. It was 102 degrees in Albany and 103 in Troy, New York, with thermometers in the sun running all the way up to 120 degrees, and bursting like Roman candles. On the Fourth of July, more than fifty people died from the extreme heat, and hundreds more were felled from prostration in what was described as the "hottest night ever" in any city in the United States. The overnight low was a blazing eighty-eight degrees. In Nashua, New Hampshire, an elderly weaver was arrested as he sat huddled before a blaze he had set in the middle of the street. With a thick overcoat sheltering him from the sun, he explained that the fire had been set to ward off mosquitoes. He was presumed insane. Boston was sizzling in 102-degree heat, Cleveland had 104 degrees, and Detroit reported 109 degrees at street level, where official readings were taken. Most cities collected their temperature data from high atop a central building where temperatures were liable to be ten degrees lower than at street level. One report observed, "Even the official temperature, which everyone wants to hear about, was discarded for the real street temperature under which he suffers."

One man, studying the foamy depths of his beer, was startled by the swinging doors of the tavern, the obvious effect of a hurried entry, to find a dog sitting up from the table next to him barking orders to the bartender. The innkeeper obligingly tossed the dog a cube of ice which he promptly left with, his tail wagging appreciation as he disappeared around the corner. Temperatures climbed to 106 degrees in Topeka and 109 degrees in Junction City, Kansas, usually the hot spot of that state. In fact, temperatures topped 100 degrees that Independence Day in nearly every city east of the Continental Divide. And the Philadelphia Athletics were sweeping four in a row from the Yankees that weekend, to knock them out of first place.

The heat and humidity was taking an awesome toll. Newspapers devoted column upon column, page upon page, listing their dead and prostrated. The clang of ambulances filled the city streets with the weather's victims, and several suicides were reported by heat-crazed individuals. Chicago broke another heat record when temperatures failed to fall below 90 degrees overnight on the fifth of the month. Parched crops were the rule throughout the grain states of Iowa, Kansas, and Nebraska, while corn and soybean crops failed in the Midwest. It was 108 degrees inside the offices of the Federal Building in Chicago, and the mercury in Junction City rose to an astounding 114 degrees. No one was in any condition to bear this weather. "Few were able to eat breakfast," and the faces of all endured the "great physical strain" under which they had been placed. Horses in the fields were dropping like flies.

No one except Colonel Atwood, who had been brightening things up with a swoop over Manhattan on his way to Atlantic City. Things caught up with him then. On the take-off of his scheduled flight from that resort town to Washington (A.C. to D.C.?), a large bull terrier dashed across the New Jersey sands just moments after his craft lifted off from the beach. The dog plowed directly into Atwood's propellers, splintering them and killing himself. These were the dog days of the summer of 1911. After making repairs, Atwood and his companion Charles Hamilton tried again, rising 200 feet into the air above the assembled crowd before diving into the ocean when their engine failed. Drenched from his unexpected bath, he turned to his companion and said, "Well, Charley, I guess we'll have to box her up and go to Washington on the train." He added, looking ruefully at his flying machine, "You're not a real aviator until you've had at least one spill anyhow."

By the eighth, the heat had broken somewhat and Atwood and his companion were ready to try again with a similar model "autoed" in from Connecticut. The craft had even passed down Broadway on its way to the Staten Island Ferry, but when the light plane lifted off this time, it sputtered and crashed into the sand, a total wreck. The heat resumed for another few days on the ninth, soaring well into the nineties, when yet another try was launched by Atwood and his companion, who had rebuilt the craft from the wreckage of the two planes. They managed to fly into Baltimore on the tenth, but the engine became sluggish from the heat, and there the flight was terminated. On July 11, believing they had made it within the city limits of D.C., they landed just before the apparent demise of the craft forced another dangerous landing. When it became known that they had fallen short of their goal, Charley Hamilton, the owner of the craft, refused to risk things any further, since the *Times* was going to award them the cup anyhow. Undaunted, Atwood bought the plane from Hamilton for $3,725.00 cash. "Now, I will go to Washington," he commented with a boyish grin.

The combination of heat and high humidities in that two-week spell claimed more than a thousand lives nationwide, while forest fires raged in Ontario, killing several hundred miners who were trapped underground. When the sun set on the evening of July fourteenth, a new Canadian air mass, albeit laden with smoke, provided a glorious background resembling Indian summer. The air was fresh and cool, and people moved more spiritedly, having endured one of the cruelest midsummer heat waves on record.

32. THE GREAT HEAT WAVE AND DROUGHT OF 1936

THE CRUEL HEAT AND DROUGHT WHICH PLAGUED NEARLY ALL THE COUNTRY THROUGHOUT the first six years of the 1930s was also a study in the stark contrasts of eastern and western farming methods. The so-called dry land farmer in the western states was subject to boom-and-bust years, while his eastern counterpart, by resorting to crop rotation, prudent use of the land, and irrigation, managed to weather the horridly hot and dry summer of 1936.

Actually, bad times began as early as 1930 in the American Southwest, and every other year brought progressively harsher conditions, with the trend for the drought to move east and northeast, touching portions of New England in the summer of '36. Nearly one half a billion dollars had been appropriated by Congress in the late summer of 1934 to bail out the indigent western farmer, and to prevent famine. Corn, wheat, and other grain crops were nearly a total failure in that year, and one by one, farmers began to leave the country for jobs in the cities. Mile by mile, the desert that had once been fertile land swept eastward from the Pacific Northwest on hot "red winds" of summer.

Some of the hardest-hit areas in 1936 included the Dakotas, Montana, Wyoming, most of the Mississippi and Ohio Valleys of the Midwest, and to the south as far as Texas. There were also an abundance of pests owing to the extremely dry conditions, and what the blazing hot sun didn't reduce to parched rubble, the boll weevil and grasshopper did. Temperatures reached into the 110-to-120 range in the first week of July, and soared to even higher levels during the balance of the month. July of 1936 established more all-time-high readings, state by state, than any other month since the Weather Bureau had been established. These factors, combined with the growing effects of the depression, ruined more lives in the prairie country than any other event before or since.

The WPA (Work Programs Administration) began to pay farmers to create ponds, terrace the land, and dam streams in order to conserve the water and stem the destructive forces of erosion. Additionally, the hard-pressed farmer was paid to build roads from the rural farm to the major market centers. But for most, this was not enough. One rancher remarked, in a kind of sustaining grim humor, "Well, the wind blew the ranch plumb into Old Mexico, but we ain't lost everything. We get to keep the mortgage." By the sixth of July, temperatures had soared to 120 degrees in Mishek, North Dakota, 119 degrees in Fort Yates, and 115 degrees in Mitchell, South Dakota. The heat had also blanketed the Midwest as witnessed by

Black Sunday dust storm bears down on Stratford, Texas, on April 14, 1935. Despite face masks, many died from suffocation. And 1936 was no better. (*New York Public Library*)

a torrid 118 degrees in Des Plaines, Illinois, and in Hell, Michigan, where the mercury stood at 108 degrees. There were, in fact, many places where it was hotter than Hell, but Hell had frozen over the previous winter, one of the coldest ever in the Michigan peninsula.

There were, in fact, more than thirty major cities throughout the western and midwestern states reporting temperatures in excess of 100 degrees, and the death toll was rising into the hundreds. While the heat records melted away, highways were reported "blowing up" from the extreme temperatures, and farmers, whose feed crops were hopelessly ruined, took their livestock to market or moved them to distant pastures. By the eighth, sporadic thundershowers eased things in the West, but a severe hailstorm in Mattoon, Illinois, beat the already withered crops into the ground. The same fields, scorched brown by the summer sun, were piled so deep in January snow that trains could not push through. When the weatherman was queried as to how the temperatures could set an all-time "high jump" record (by ascending from an all-time wintry low of minus fifty-five degrees at International Falls to 120 degrees across the border in Mishek) he shrugged, "The wind shifted."

On the ninth of July, the blanket of heat invaded the East as temperatures in New York City registered an official 106 degrees, 415 feet above street level, while in Times Square and the canyons of Wall Street, it was 115 degrees. The official reading in the sun was 145°F, from searing blasts of heat borne out of the parched drought areas of the Midwest. On the evening of the ninth, it was still 100 degrees on the street after sunset, and an astronomet toyfully pointed out to his languid audience that it was 750°F on the planet Mercury, and if this wasn't enough to cool them off, it was -250°F on Saturn. In the town of Two Rivers, Wisconsin, a four-foot-thick snowbank was uncovered in the midst of the 100-degree heat.

By the tenth, the blistering heat wave that cut a path of death and destruction through the Plain states and Midwest, continued to roll eastward smothering city and town, hamlet, and farm under a hot blanket of devastation. Thousands of dry-land farmers who tried to raise a crop on land that never should have been put to a plow reached the end of the trail, leaving the parched country for other states. Aban-

Effects of the dust bowl. (*National Archives*)

doned farms and buildings, with roofs beginning to sag, unpainted for many years, appeared extensively throughout the Plains as stark reminders that the range country cannot be farmed intensively without costly irrigation.

On the eleventh, rains came to the Northwest and Midwest, and people were seen running from their homes to stand in the long-prayed-for showers that temporarily dropped temperatures to the sixty-degree mark after nine straight days over 100 degrees. The following day, the heat resumed as the torrid rays of the sun, blazing down relentlessly, prostrated hundreds more as the nation's death toll neared the 1,000 mark. The sudden rains of the previous day were, in large, wasted as the droplets splattered off the denuded land, offering little resistance to the wasteful overflow. By the end of the second week of July, the nation's death toll rose above 3,000, and damage to crops was estimated to be in excess of one billion dollars.

For the balance of the month, temperatures eased off a bit in the East, while the West baked in sizzling 100-degree heat and the shortage of water for human consumption reached the critical stage. New wells were forged and old ones deepened in the frantic search for life-bearing water. President Roosevelt reiterated the common feeling when he said, "The fact is, we have been prodigal with our land and we shall suffer for that prodigality until we have restored what has been wasted." Felix Belair, a correspondent for the New York *Times*, wrote, "The largest part of the Spring wheat area is burned to a crisp. Montana and Dakota farms and many in Minnesota are now pictures of desolation that lifetime tillers of the soil say has never been witnessed before. Fields that at this time last year were waving with tall green and yellow grain now appear dark brown and black. The range country seems covered with a tan moss so close to the ground that hungry cattle cannot reach it. So dry is the covering that it is useless for sheep. Grasshoppers by the millions seem bent on outdoing the drought as the destroyer of crops are everywhere...when one field is finished, they fly in droves to another." Even travelers by auto were seen to stop occasionally to scrape their battered bodies from their windshields and radiators.

There is no certain way to predict drought, other than a loose relationship between the twenty-two-year sunspot cycle which corresponds, in some areas, to wet and dry cycles of the weather. When the Weather Bureau announced at the end of July that "very little pasture is now available between the Rocky and Appalachian Mountains," they summed up a disturbing aspect of life in the middle latitudes of the globe. For the heat wave and drought which took shape in the spring of 1936 and stretched across the country in a broad belt from a little west of central Montana to Kentucky, Tennessee, and South Carolina and northward on into southern New England could not be prevented. The high temperatures and the persistently hot winds evaporated what little soil moisture was available and sapped a nation's water resources to a calamitous dimension. In the East, the situation was alleviated somewhat by the planting of trees on hillsides and groundcover to soak up the "little waters," but in the western United States the blast of the drought became a monster too great for man to grapple with. More than five thousand fatalities were directly attributed to the forces of Nature in that grim summer, and, unlike the direct assault of a flood or hurricane, its effects, though subtle, were as devastating as any natural disaster that has occurred in this country in the last 150 years.

33. A WARTIME HEAT WAVE

IT WAS THE SUMMER OF 1944. U.S. TROOPS, LED BY GENERAL DWIGHT EISENHOWER, HAD already established a beachhead at Normandy, France, in the spring, and the choice of D-Day was determined on the basis of there being enough moonlight for airborne operations, low tide on the beaches so that the demolition teams could see and destroy the numerous underwater obstacles, and sunrise at the right hour to allow enough time for one final air and naval bombardment before the first wave of troops hit the beach. The general state of the wind and the weather were of critical importance. June 5, 1944, was selected as D-Day, based on the foregoing criteria, but Nature stepped in with a dismal wind and rainstorm. General Eisenhower postponed the operation until the next day, and on that stormy morning the critical second front in Europe became a reality.

This is yet another example of one of the several major battles that have been decided by the weather. Weather played a key role in the Revolutionary War in 1776 when General George Washington was surrounded by Cornwallis's troops at Valley Forge and was mired in the mud. Recognizing that a cold front had passed on Christmas Eve of that year which would freeze the ground sufficiently for his troops to decamp under cover of night, they lit campfires to belie their presence and escaped certain demise with the benefit of the north wind.

Weather, meanwhile, was making the news in the United States. Hot air was building up in the southern states in early June and was destined to make the summer of '44 the hottest since 1896. Beginning with the last week of June, and extending through mid-July, temperatures east of the Rockies, and specifically east of the Mississippi River, ranged well above normal every single day. Readings throughout the Midwest and South topped 100 degrees, while the normally more humid and oppressive Atlantic coast broiled well into the 90s. After a week's respite, temperatures again resumed their upward climb, peaking out during the ten-day period between August tenth and the nineteenth, as General Patton was organizing Allied troops in southern France, on their way to the invasion of Germany.

New temperature highs throughout the eastern United States had already been established by the scorching heat in the early part of the month, but the worst was yet to come. By the eleventh, naturalists exploring the Bear Mountain State Park, about thirty miles up the Hudson from New York City, discovered that the combination of drought and heat had taken its toll on the park's wildlife. Nearly all the rivers and streams which flow through the park had gone dry, and nearly all of the park's wildlife, animals, and birds alike were clustered around the park's thirty-two beaver ponds, which offered the only green foliage and water. While the deer, fox, and raccoon mingled with the heron in Bear Mountain, forest fires broke out in many eastern parks, forcing many hikers and campers out of the woods for the rest of the summer.

By the fourteenth of August, widespread ice shortages plagued homeowners, theaters, and taprooms, as the effects of the heat diminished supplies to a point where only hospitals and dairies were served. The heat had scorched the wooden structures at Palisades Amusement Park to a point of brittle tinderwood, and as sparks from the Virginia Reel, an all-metal ride, ignited a nearby structure, the fifteen-acre park soon went up in flames. Nearly 100 automobiles caught fire and exploded like bombs bursting in air, and swimmers were forced by the expanding conflagration to evacuate the pool. Nearly 30,000 people were routed from the park, many with only bathing suits or towels for cover. As the usual exits were barred by flames which shot up more than 100 feet into the air, they made their escape through underground tunnels. For some reason, the ferris wheel was left running and a mother was seen grabbing her son, whose clothes were afire when his cab reached the bottom. Palisades Park was not the best place to cool off on that extremely hot day. It was the third major fire in as many days in the New York metropolitan area.

Over the following two days, violent thunderstorms pounded various areas of the Northeast without offering very much by way of relief. For example, one such storm blackened the skies over Gotham on the evening of the sixteenth and dropped temperatures from 90 degrees to 67 degrees in only one hour, while dumping from two to three inches of subway-flooding rains. Nearby Newark Airport reported more than six inches of rain, accompanied by a steady bombardment of lightning and wind gusts surpassing sixty-five miles an hour. After the storm passed, temperatures began to rise once more, and it was not until the nineteenth of the month that the back of the searing heat was broken.

Near the close of the long hot summer of '44, a hurricane packing winds of 140 miles an hour lay off the south Atlantic coast poised for a destructive landfall near Cape Hatteras and all shore points north to the Canadian border. The evening of September thirteenth and the day of the fourteenth were wild ones for East Coast residents, who were lashed by drenching rains and winds which crested at 100 miles an hour in Atlantic City, where their famous Boardwalk was completely demolished by waves estimated to be fifty feet high. Perhaps the most interesting feature of this hurricane was that it was the first to be penetrated by reconnaissance planes, whose findings overturned previous assumptions about the nature of tropical storms. One fact that was revealed as a result of the daring flight undertaken by Colonel Lloyd B. Woods was that a hurricane had ascending winds within its inner edges, while the air descended sharply around the perimeter of the storm. His account follows:

> For the first 50-60 miles, we encountered strong down currents. There was heavy rain and black clouds. Occasionally, we could see ocean spray which must have been 50 to 60 feet above the sea's surface. We flew about 100 miles further along and as we got closer to the center, there was a strong rising current, about 2,000 feet per second.

The plane encountered three distinct motions—a forward speed of 300 miles an hour, a sheering sideward motion from the counterclockwise winds of the tempest of 100 miles an hour, as well as the upward thrust experienced near the eye.

The summer of 1944 was a turning point in the war for the Allied forces, a time of discovery off the Mid-Atlantic coast, and one of the most oppressive hot-weather seasons of the twentieth century for millions of residents in the eastern half of the nation.

34.
IN THE HEAT OF BATTLE

THE SUMMER OF 1948 FOUND THE NATION IN THE GRIPS OF ONE OF THE MOST EXCITING pennant races of all time and an equally close presidential contest. Fortunately for the contemporary American, the barrage of campaign rhetoric didn't get under way until nearly Labor Day, but convective blasts of hot air from other sources did. In New York City, for example, the summer of 1948 provided the only example of three back-to-back days with temperatures of 100 degrees or better in the more than 110-year period the Weather Service has been in existence.

The heat wave in question originated in the American Southwest, with temperatures in Tucson and Phoenix cresting over 110 for several days with the commencement of summer. Gradually, the hot dry air drifted eastward and northward, causing temperatures throughout the East to ascend over the century mark by June twenty-seventh. An interesting, albeit gruesome story emerged from rustic Fairfield, Maine. Carl Fisher, who worked in the stockyards, returned home late on the night of June twenty-ninth to find his three children, aged six, four years, and nine months, drowned by his wife who was semi-comatose as result of a drug overdose. A note read, "I'm sorry I have to do this. I haven't done a proper job of raising the children. They will be better off in heaven." After five years of rehabilitation, Mrs. Fisher, ruled insane by the courts at her trial, reemerged to again rear three children by her former husband, and in the sordid heat wave of June 1966 drowned her second three children in a similar fashion.

Temperatures remained in the nineties for the balance of June, and July was abnormally hot in the eastern two-thirds of the nation. Ethel Merman was starring in *Annie Get Your Gun*, and the Brooklyn Dodgers had just pulled off a triple steal led by Jackie Robinson's fourth steal of home of the season. The Brooklyn Bums, who had been in the cellar as recently as the second of July, now found themselves in first place after an awesome midsummer drive. By the twenty-third of August, temperatures in Chicago soared to the century mark, and in Cleveland, the ageless black pitcher Satchel Paige threw a three-hitter for the front-running Indians. While the Indians were soon to flounder, the temperatures in Cleveland did not. Temperatures there passed the hundred-degree mark for three days straight, and then the heat moved east.

Hartford, Connecticut, reported an astounding 105 degrees on the twenty-sixth, while 100 degrees or more was attained everywhere from Charleston, South Carolina, to Portland, Maine.

Meanwhile, off the Florida peninsula, a hurricane was brewing which was to be dubbed Harry after the incumbent president. Packing 120-mile-an-hour winds, it moved up the coast to a menacing point about 150 miles east of Cape Hatteras before blowing out to sea and taking the heat wave with it by the end of the month.

September 1948 brought a general cooling off of the weather, but three hurricanes threatened the East Coast in a very active season. Meanwhile the pennant race heated up. Cleveland had been knocked out of first by the resurging Boston Red Sox and the perennial contenders, the New York Yankees. It was nip and tuck in the latter days of August and the entire month of September. The American League title was not even decided on the final day as three teams went to the wire in a virtual deadlock, with Cleveland and Boston winding up with identical records. Meanwhile, in the National League, Boston's pride, the Braves, had blown out the Dodgers as contenders and wrapped up the pennant with two weeks to spare.

In a one-game play-off, Cleveland routed Boston in the New England City and went on to astound the Braves in a six-game series led by manager-player Lou Boudreau. Funny how the baseball season spans the entire hot weather season and lingers well into the hurricane season. While pedestrians wilt and pavements buckle, the national pastime just rolls along, oblivious to all but the worst possible conditions. 1948 was the year of the Alger Hiss-Chambers debates before the House Un-American Committee, and a strong anti-Communist passion swept the nation and nearly put Thomas Dewey into the White House. Shortly after Truman had had a hurricane named after him, a stronger more intense disturbance dubbed Bess dumped more than nine inches of rain on the already drenched palm trees of Florida. That hurricane also moved up the East Coast to damage the incoming Queen Mary with the forty-five-foot waves it generated.

1948 turned out to be the year of the underdog, and it's surprising that the third party candidacy of Henry Wallace, the peace candidate, did not succeed. There are a few instances of back-to-back 100-degree days recorded in the annals of most eastern cities, but 1948 stands out as the only summer producing three or more in a row. History also shows a strong correlation of very hot summers being followed by numerous hurricanes along the East Coast and, as always, thrilling pennant races. A prominent sportswriter was asked, two weeks into September, who he thought would clinch the flag, and he answered in all candor, "You got me, bud." He sounded quite a bit like the weatherman responding to a query about when the heat wave of 1948 would end. Modern pundits of the weather and baseball and presidential politics should take note.

35.
AN UNUSUALLY EARLY HOT SPELL

VERY SELDOM DO TWO BACK-TO-BACK YEARS COME ALONG WHEN THE RECORD BOOKS ARE rewritten to the extent that they were in 1976-77. Who can forget the brutal cold that persisted from the fall of 1976 well into the winter months, forcing plant and school shutdowns and a widespread shortage of natural gas? The record books are generously sprinkled with record-setting low temperatures for the entire period encompassing the middle and late seventies—the latter years producing impressive snowfalls to compound matters. Ironically, these years possess an abnormally high percentage of record-high temperatures.

The winter days of the 1975-76 came to an abrupt end in mid-February as temperatures soared nearly fifteen degrees above normal in the final two weeks. It had been an open winter in most eastern states, and early flowers and shrubs burst forth ahead of schedule. This pattern persisted into March, but it was not until the Easter celebration of April that some spectacularly summerish highs were established. It was the beginning of roughly three years of wild temperature and snowfall extremes that most states east of the Mississippi were about to experience.

By the fifteenth of April, temperatures were blossoming into the eighties in the midwestern cities of Chicago, Detroit, and Cleveland, and the warmth was rapidly spreading eastward. The baseball season opened with Florida-like temperatures, and the Yankees christened their renovated stadium with a victory, exactly fifty-six years after Babe Ruth smacked a home run on his first appearance in the House that Ruth Built. The following day, Good Friday, heavy snows fell in New Mexico and Arizona, but the East was basking in summerlike weather as readings soared all the way to eighty-eight degrees in Philadelphia and ninety-three in Richmond. On Saturday, temperatures surpassed ninety degrees in most eastern cities, setting numerous records for such an early-season feat. Noteworthy among other early-season events was Mike Schmidt's performance for the Phillies that day. He walloped four consecutive home runs, a feat that has not been duplicated this century in either league, as the Phils beat the Cubs 18-16 in ten innings in the season opener.

It was unusual to see dogwood trees and azalea in full bloom so early in April as far north as New England, but by Easter Sunday, the eighteenth, many record-high readings for the month of April as a whole were to be surpassed. The mercury surged well into the nineties, and all-time highs for the month toppled in Boston, New York, Philadelphia, and Washington, and several people plunged into the surf as far north as Cape Cod, where ocean temperatures stood in the middle forties. I recall taking the plunge myself, in a stream in Connecticut, where the icy chill of the water soon evaporated in the blazing sunshine.

On Monday, the Boston Marathon was run in the hottest conditions ever in the history of the event as temperatures surged past ninety degrees for the third straight day. Farmers in upstate New York watched in dismay as their summer crops burst forth, for the farmers were aware that frosts occurred as late as the end of May, and it was frightening to see changes which normally took several weeks to unfold happening in a matter of a few days. The heat persisted until midweek, when relief arrived, but the high temperature for the year had already been established, hotter than July in April.

Ironically, the summer of 1976 produced little by way of weather news, but in the autumn, many new low readings were set in the East and persisted until March and April of 1977, when many more temperature highs, some of which were set only the year before, fell by the wayside. In fact, the spring of 1977 produced the warmest March days of all time and continued into April twelfth, when most eastern cities recorded their earliest ninety-degree-plus day. The lesson seems to be that wild weather conditions beget their mirror image, but never was this lesson so stark as it was in the spring seasons of 1976 and 1977.

36.
THE GREAT SOUTHWESTERN HEAT WAVE OF 1980

SHORTLY AFTER THE CONCLUSION OF THE POLITICAL PRIMARY SEASON OF 1980, AND UNDER the spotlight of a foregone nomination process, many weather-conscious eyes in America turned their attention to the Southwest, particularly to the Dallas-Ft. Worth area. What had caught so much media attention was the string of consecutive 100-degree-plus days being compiled there. Beginning in the second week of June and lasting through much of August, with the exception of a brief respite afforded by the passage of hurricane Allen, temperatures were the hottest known for such an extended period in the heart of Texas and for much of the country's breadbasket in general.

The usual concomitant of high heat—drought—accompanied the scorching temperatures throughout the summer months and extended well into Arkansas, Kansas, Missouri, and Illinois, forming the core of the torrid weather. Many records for individual high readings and for duration fell in the summer of 1980, while these areas were receiving only twenty-five percent of expected rainfall amounts. The combination of heat and drought brought economic disaster to much of this region, as crop losses numbered twenty billion dollars and the human toll stood at 1,200 by summer's end.

The heat wave became established on June tenth when El Paso began a string of days in which the unobscured sun baked the area with one-hundred-degree readings and scarcely a drop of rain fell for the entire month. Two weeks later, after the passage of the summer's only cool front, Dallas began its incredible series of forty-two days over the century mark. As early as the twenty-sixth and twenty-seventh of June, Dallas recorded back-to-back all-time temperature marks when the mercury ascended to 113°F. High pressure over the American Southwest allowed the sun to scorch the land, day after day and, during the month of July, Dallas reported a new record *average* mark of ninety-two degrees. This was a result of daytime maximums over the century mark, with early morning lows only about eighty degrees.

By the middle of July, the scourge of summer heat and scanty precipitation moved eastward to affect the humid southeastern states, the Mid-Atlantic, and New England. Similarly, after the dominant high-pressure system moved back to a more westerly position centered over the four corners of New Mexico, Utah, Arizona, and Colorado, much of the Pacific-states region felt the brunt of the heat and persisting drought. During this period, new all-time-maximum readings were attained in the southeastern cities of Macon,

Georgia (108°F), Meridian (107°F), Pensacola (106°F), and Atlanta (104°F). Temperatures crossed the century mark at New York City, Washington, D.C., Baltimore, Maryland, and many other locations in between, on the twentieth of July. The nation's Capitol eclipsed the former record in July, for the all-time-hottest month in its history, averaging a steamy 82.3 degrees.

Meanwhile, out west, the fertile valleys of Sacramento were drying up in the final two weeks of July, as the daily average maximum reached a stifling 107 degrees. Even the normally cooler northwestern cities of Portland and Boise eclipsed the one hundred mark on the twenty-seventh of the month. The drought continued to make itself felt over a broad belt extending from the wheatlands of the Northern Plains, southeastward into central Texas, and then east and northeastward across a significant part of the country's breadbasket and to nearly every coastal state in the East.

It was not until the fourth day of August that Dallas's impressive run of scorching conditions retreated temporarily, ending that record string. It was on that day that Hurricane Allen earned itself a name as it churned through the South Atlantic. Following its passage through the Virgin Islands, where more than two hundred lives were claimed, it entered the Gulf of Mexico and became the second most intense hurricane in that area's history with a central pressure of 990 millibars and reported wind gusts of 195 miles an hour near its central core. Due to these spectacular figures, the southern states looked warily at the storm both as a potential disaster or, alternately, as the long-awaited relief from heat and drought. Traveling at a forward speed of eighteen miles an hour, hurricane Allen pursued a direct westerly course and smashed into the lower-Texas coast between Brownsville and Corpus Cristi, in the underpopulated vicinity of Padre Island. Winds of 165 miles an hour accompanied the eye of the storm, and rainfall amounts throughout the southern half of Texas amounted to between fifteen and twenty inches. This copious moisture, although much needed, was more than necessary as drought turned to flood. Similarly, some much-needed rain fell in the Midwest, averting a major crop disaster there.

A major summer wave developed, affecting both eastern and western sections of the country, in the unusual summer months of 1980. Ordinarily, either the West suffers while the East is relieved by dry cool Canadian air masses, or the situation is reversed, but it is rare for the entire country to be under such a canopy of oppressive weather. The highest reading observed in the United States in 1980, 117 degrees, occurred at Wichita Falls, Texas, on the twenty-eighth of June, while many other stations came close to this mark. Similarly, new records for precipitation were established in Annette, Arkansas, Richmond and Lynchburg, Virginia, for the driest June. Each of these reporting stations measured less than an inch of rain for the entire month. Temperatures in July reached 100 degrees or better in each of the contiguous forty-eight states except for Michigan, West Virginia, and parts of New England. It was not until August that rainfall patterns and temperatures resumed more moderate ways, ironically touching off one of the wettest eighth months in parts of Ohio, West Virginia, and southwestern Pennsylvania. These renewed rains both saved the economic plunder of the Corn Belt, while yielding flash floods to the areas that had been praying for rain in July. It goes to show, as the rock group the Rolling Stones prophesied in one of their recordings, "You can't always get what you want...but if you try, sometimes you get what you need."

37. HOW THE BODY'S HEAT MECHANISMS WORK AND HEAT WAVE SAFETY RULES

MAMMALS AND BIRDS ARE HOMOIOTHERMS—WARM-BLOODED CREATURES THAT MAINTAIN an essentially constant body temperature regardless of their thermal environment.

Human bodies dissipate heat by varying the rate and depth of blood circulation, by losing water through the skin and sweat glands, and, as the last extremity is reached, by panting. Under normal conditions, these reflex activities are kept in balance and controlled by the brain's hypothalamus, a comparatively simple sensor of rising and falling environmental temperatures, but a sophisticated manager of temperature inside.

A surge of blood is heated above 98.6° F sends the hypothalamus into action. The heart begins to pump more blood, blood vessels dilate to accommodate the increased flow, and the bundles of tiny capillaries threading through the upper layers of the skin's surface are put into operation. The body's blood is circulated closer to the skin's surface, and excess heat drains off into the cooler atmosphere. At the same time, water diffuses through the skin as insensible perspiration, so-called because it evaporates before it becomes visible, and the skin seems dry to the touch.

If the hypothalamus continues to sense overheating, it calls upon the millions of sweat glands which perforate the outer layer of our skin. These tiny glands can shed great quantities of water (and heat) in what is called sensible perspiration, or sweating. Between sweating and insensible perspiration, the skin handles about ninety percent of the body's heat-dissipation function.

As environmental temperature approaches normal body temperature, physical discomfort is replaced by physical danger. The body loses its ability to get rid of heat through the circulatory system because there is no heat-drawing drop in temperature between the skin and the surrounding air. At this point, the skin's elimination of heat by sweating becomes virtually the only means of maintaining constant temperature. Now it is not the heat but the humidity, as they say.

Sweating, by itself, does nothing to cool the body, unless the water is removed by evaporation—and high relative humidity retards evaporation. Under conditions of high temperatures (above 90° F, or 32° C) and high relative humidity (above seventy-five percent), the body is doing everything it can to maintain 98.6 degrees inside. The heart is pumping a torrent of blood through dilated circulatory vessels; the sweat glands are pouring liquids—and essential dissolved chemicals, like sodium and chloride—onto the surface of the skin. And the body's metabolic heat production goes on, down to the vital organs.

Reprinted with permission from *Weatherwise*, June 1980, Volume 33, Number 3, page 115

HEAT WAVE SAFETY RULES

1. **Slow down**. Your body can't do its best in high temperatures and humidity, and might do its worst.
2. **Heed your body's early warnings that heat syndrome is on the way**. Reduce your level of activity immediately and get to a cooler environment.
3. **Dress for summer**. Lightweight, light-colored clothing reflects heat and sunlight and helps your thermoregulatory system maintain normal body temperature.
4. **Put less fuel in your inner fires**. Foods (like proteins) that increase metabolic heat production also increase water loss.
5. **Don't dry out**. Heat wave weather can wring you out before you know it. Drink plenty of water while the hot spell lasts.
6. **Stay salty**. Unless you're on a salt-restricted diet, increase your salt intake when you've worked up a heavy sweat. (But remember that the American diet tends to be too heavy on salt; many people can get adequate salt from their normal diet, or by slightly increasing their salt intake.)
7. **Avoid thermal shock**. Acclimatize yourself gradually to warmer weather. Treat yourself extra gently for those first two or three critical days.
8. **Vary your thermal environment**. Physical stress increases with exposure time in heat-wave weather. Try to get out of the heat for at least a few hours each day. If you can't do this at home, drop in on a cool store, restaurant or theater—anything to keep your exposure-time down.
9. **Don't get too much sun**. Sunburn makes the job of heat dissipation that much more difficult.

Courtesy National Atmospheric and Oceanic Administration

VII. HURRICANES

The hurricane-prone coast of the United States extends thousands of miles from Brownsville, Texas, in the western Gulf of Mexico to Eastport, Maine, along the north Atlantic. In the period from 1871 to 1973, 281 tropical storms have affected the Gulf coast of the United States while 131 have buffeted the Atlantic coast, for a total of 412. Ironically, nearly thirty million people have moved into these vulnerable areas in the past twenty-five years, and most have never experienced a hurricane of any dimension. One reason is that in the past twenty years, since the passage of Donna in 1960, hurricane activity has been on the decline although the long-term averages suggest that nine tropical storms will develop—five of them of hurricane force—in any given year.

Hurricanes are noted in the Atlantic as far back in time as Columbus's fourth and final voyage to the New World. In the summer of 1502, he saw the early warning signs and requested shelter in the Santo Domingo Harbor, off the island of Hispaniola, but was refused. Thirty ships loaded with gold and slaves set out for Spain, never to be seen again. Columbus wisely rode out the storm at anchor in another island cove and survived. In July, 1609, a fleet of ships crowded with settlers bound for the Virginia colonies were overwhelmed by a tropical storm near the island of Bermuda. Ten months later, the passengers of the flagship Sea Adventure *limped into Jamestown in a small boat built from the wreckage of their former vessel. Three years later, upon hearing their tale, William Shakespeare was to write the moving story* The Tempest. *On November 2, 1743, a hurricane passed near enough to Philadelphia to obscure the expected eclipse of the moon. A disappointed Ben Franklin wrote to a friend in Boston about the event and was surprised to hear that the eclipse was magnificent in New England although they had a "terrific blow" the following day. Based upon this information, Franklin deduced the modern concept that all storms have a circular wind motion and that a storm was not necessarily coming from the direction of the surface winds.*

The official hurricane season runs from the first of June until the end of November, although tropical storms have been active in the Atlantic as early as March seventh (1908) and as late as

December thirty-first in the memorable season of 1954. The earliest hurricane to actually strike the U.S. mainland was Alma on June 9, 1966, while the latest was an unnamed tempest which hit near Tampa, Florida, on November 30, 1925. While most tropical storms are active for a matter of ten days to three weeks, Ginger was a hurricane for twenty days and a full-fledged tropical storm for thirty-one.

Early and later hurricanes are usually spawned in the Caribbean while those in the core of the season—from about the middle of August to the first of October—frequently are born thousands of miles away in the Lesser Antilles, near the western shores of Africa, and take nearly two weeks to cross the Atlantic. While the mechanics of a hurricane's development are not fully understood, it is known that they require the proper amount of heat (with surface water temperatures above 80° F.), moisture, and atmospheric instability. It is the earth's rotation, or Coriolis Effect, that gives them the necessary "swing" to generate their circulation around a calm center. The eye of the storm is similar to a chimney, and the air is much warmer in this central core than in the surrounding atmosphere. The warm air ascends, creating thick bands of clouds frequently as high as eight miles, touching the lower limits of the stratosphere. Rains develop, winds increase, and a hurricane is born. It is the prevailing winds of the upper atmosphere that guide a hurricane on its often erratic course. Generally, they drift with the prevailing easterlies of the lower latitudes and begin to curve to the north when they reach the twenty-fifth parallel. Over the period of their lives, winds can increase to 200 miles an hour or more, and rainfall amounts exceed five to ten inches. The record rains from a hurricane in twenty-four hours is forty-six inches!

Tropical disturbances are classified from a range of one to five, with five being the most intense. Known as the Saffir/Simpson hurricane scale, there have been only sixteen hurricanes this century rated as a four or five, several of them being discussed in this chapter. In order to rate a five, which only three hurricanes this century have accomplished, winds must be in excess of 155 miles an hour, barometric pressure less than 27.17 inches (920 millibars), with the accompanying storm surge greater than eighteen feet. The most intense hurricane of the century struck the Florida Keys in 1935 with a record low barometric reading of 26.35 inches (892 millibars).

While a storm's intensity is a reliable gauge of the winds, it does not necessarily make it deadly or costly. The costliest storm before Frederick in 1979 was Agnes in June of 1972, rated only a one on the hurricane scale, but the attendant flooding in densely populated areas of the Northeast claimed most of the toll. Similarly, Diane of the 1955 hurricane season rated only a one but proved to be near the top in the human and property tolls. The Galveston storm of 1900, which killed upwards of 12,000 people and ranks number one in death toll, was only a four, but it was the tremendous storm tide estimated to be fifteen feet or more which claimed most of the lives.

It is the major population centers which have developed along America's coastline in the past twenty to thirty years, a quiescent period for hurricanes, that are vulnerable to major disaster in the future. Many shoreline communities are built on sand, and even with the sophisticated advance-warning systems weathermen have at their disposal today, massive evacuation plans for these communities have yet to be developed. One further concern is the relative nonchalance that many regard hurricanes, and this can lead to the notion of "riding out" a hurricane, as few have adequate respect for the fury of these storms.

One final note on the naming of hurricanes. It was not until the early fifties that an effort was made to distinguish one from another, and World War II meteorologists were in the practice of naming them after their girl friends. Thus, an all-girl list was composed, and this system remained in effect until 1979, when five lists of girl/boy names was devised and approved by an international commission (The World Meteorological Organization). There will be a unique list of names until 1984, when the 1979 list will be reused, completing the five-year cycle.

38. A VINTAGE YEAR FOR HURRICANES—1893

NO FEELING OF REGRET ATTENDED THE DEPARTURE OF THE SEASON'S LAST TROPICAL STORM on November 15, 1893. No season to date had consistently battered the United State's East Coast with cyclone after cyclone nor caused so much widespread devastation and death. In fact, on August twenty-second of that year, four hurricanes co-existed in the Atlantic, one of them being the hurricane which killed an estimated 2,000 people in Georgia and South Carolina a few days later.

Communications in that era were only as certain as their telegraph wires were secure. It was not unusual at the close of the nineteenth century for major cities to be isolated from the world, cut off by the passing of an unusually heavy snowfall or full-scale hurricane. In fact, it happened so frequently that, as the twentieth century opened, a move was already afoot to lay underground cables. For a ten-day period, from August nineteenth through the twenty-ninth, three full-fledged hurricanes lashed at the East Coast in an unprecedented barrage of wind and wave. The first of this series passed the coast to the east, spending much of its fury at sea, from the nineteenth to the twenty-first. The second caused extensive damage from Florida to New England and claimed thousands of lives in the South and again as the eye passed just to the west of New York City on the twenty-fourth. The third hugged the coast before coming ashore near the Virginia-North Carolina border and passed through east central Pennsylvania on the twenty-ninth of the month.

Few people slept a wink the night of the twenty-third of August along the coast. In New York City, roofs were ripped off and chimneys toppled, and many of the low-lying districts were under several feet of water. During a long night, hour after hour, this hurricane poured out its energy relentlessly, and Central Park was devastated as never before. Hundreds of prize trees were torn up by their roots, while tens of thousands of birds were drowned or bludgeoned to death by the cruel winds. Enduring their second major hurricane in a week, "it looked to some timid persons as if the man that makes the weather had got his Fall stock of storms badly mixed." One measure of a storm's intensity is the barometer, and it crawled down its narrow tube to 29.23 inches, a figure exceeded on only two other occasions.

The coasts of New Jersey and Long Island were swamped by the tremendous force of the wind and tides, and shore homes were utterly destroyed just hours after evacuation. At the Hoffman House in Rockaway Beach, where an annual ball was in progress, waves pounded a 300-foot iron pier, and few of the revelers felt any danger. Suddenly the building began to tremble. A loud noise was heard above the music of the band, and within a moment's notice, a thirty-foot section of the pavilion caved in, and an angry sea poured into the building. The dancers fled in terror. Every boat within reach of the waves was smashed on the beach, and the iron pier was swept out to sea. In Long Island City, a large building which had burned in July, so there was nothing left but menacing walls, was crushed, thereby accomplishing what a large force of men with dynamite could not in the previous weeks.

As is the case in every storm of this dimension, marine disasters were numerous, and many a barge was sighted through the early morning mist, drifting aimlessly as though disabled. Most, however, were involuntarily abandoned. One tug, the *Panther*, just drifted ashore to a sandbar off South Hampton, where lifesaving crews found no sign of life. One sailor, who was insensible and lapsing into unconsciousness, related how the tug sprang a leak and filled so rapidly that there was barely enough time to adjust life jackets before everyone was swept overboard by the waves. The captain of the tug related, "I saw the black wall approaching...I knew what it meant...a towering crushing mountain of water, about to break. I screamed to the men to save themselves and at the same time grasping a rope and clinging to it with all my might...and then the huge wave broke over us. My God, but it was awful. It swept our entire deck and I was dashed against the mainmast...my mouth was filled with water and my eyes blinded by the salt foam. I thought I was drowning." Thirty-four others did drown, and all of the sailors—dead and alive—were washed up on the storm-lashed beaches, their life preservers intact.

Another ship met a similar fate. Every living soul was washed overboard, with only seven men clinging to the end of the rigging as the boat lifted its bulwarks above water. Somehow, seven of them managed to scramble back on board before another monstrous wave broke over the drowning craft. The huge, foamless swells from the south gave unmistakable notice that the Storm King was at hand. Half fainting and breathless from their struggle with the elements, five more men were unable to withstand the storm's fury, and let go only to find a watery grave deep in the sea.

Captain Peterson and his first mate, a man by the name of Cummings, staggered upon the deck only to find it otherwise deserted. The captain shouted, "Cummings, my boy, keep cool, it's our turn next." They couldn't hold on much longer. The next wave swept them into the ocean, far from the craft. The first mate cried out, "We'll make it all right," and that was the last they were to see of each other. The captain, the only survivor, said he saw others floating on planks, and he grabbed the top of a pilothouse which had been washed away. "I suddenly turned over and remember no more till I came ashore in the hands of my kind friends here."

In the following days, residents of the Northeast had no way of knowing that yet a third hurricane was churning northward from the Caribbean that was destined to kill over a thousand people, mostly blacks, on a trail of destruction from Titusville, Florida, to Wilmington, North Carolina. The storm eventually passed through Harrisburg, Pennsylvania, where again communications were silenced. Yet a fourth hurricane crossed the area over Columbus Day, again passing through central Pennsylvania and causing widespread flooding. At the end of the hurricane season of 1893, many people living along the East Coast of America saved nothing but their lives and would welcome any change, even the onset of winter.

39.
GALVESTON, 1900 — AMERICA'S WORST NATURAL DISASTER

NO WORDS CAN ADEQUATELY DESCRIBE THE HORROR THAT SURVIVORS FACED ON THE GRISLY morning of September 9, 1900, in Galveston, Texas. What an awesome amount of courage was summoned by the living to carry on and rebuild a city which lay in total ruin. Whole families were wiped out, and nearly every family lost at least one member as a result of America's deadliest hurricane and the ensuing floods. The whole story can never be told, and the death toll will never be known, but estimates range from 10-12,000 lives. Every survivor faced death on the night of September eighth as flood waters turned the city of Galveston into a raging sea.

Galveston, according to a 1900 census released only a few days before, had a population of 37,789. It was a thriving city, one of the wealthiest in the country, and was located at the foot of Galveston Island at the mouth of Galveston Bay, on the Gulf of Mexico. Galveston Island is actually a long barrier beach which parallels the Texas coastline, connected to the mainland by America's "longest wagon bridge." It is a city literally built on sand, with an average altitude of six feet above sea level. In 1872, the entire eastern end of the island had been swept away by a tidal wave that followed a terrific hurricane that swept through the gulf for three days. It's surprising, then, that a seventeen-foot seawall was not constructed until 1902.

A tropical storm had been tracked for days by the Weather Bureau as it trekked across the Caribbean, south of Puerto Rico and directly over Cuba. It did not gain hurricane status until it entered the gulf on the fifth. By the sixth, winds reached 100 miles an hour near the center, which was now poised about 600 due east of Galveston where the sky was overcast, with temperatures in the eighties. Menacing waves soon drove bathers out of the gulf, and the swells increased steadily until several blocks of the city were under water Saturday morning, the eighth. By 7:45 A.M., it began to rain, and several beach houses had already been destroyed by the thundering surf.

By noon, the northeast wind increased to thirty or forty miles an hour and it was now raining heavily. The barometer was falling rapidly and the water stood four feet deep in the streets. The wagon bridge connecting Galveston to the mainland was submerged, and people beat a hasty retreat to somewhat higher ground. It was not uncommon to find 50 or more people—neighbors, friends, and family—huddled together in what seemed to be the sturdiest house on the block. The people of Galveston were not timid, and

Relief party working at Avenue P and Tremont Street after the hurricane. (*Library of Congress*)

many could recall weathering other hurricanes from the gulf. This was nothing new.

By mid-afternoon, it became more and more evident that the city would be visited by disaster. One of the last communications to leave Galveston that day read, "Gulf rising rapidly, half the city now under water...a great loss of life must result." By 3 or 4 P.M., the waters of the bay met the gulf, and the city was entirely submerged, with winds rising constantly and rain falling in torrents. Electrical and gas plants were flooded, and by nightfall the city was plunged in darkness. As the waves ate away at their foundations, buildings caved in and crashed, and the noise of the wind was terrifying.

Suspicion began to mount that an awful calamity rested behind the lack of information from the gulf. It was nearly a week before communications with the outside world were to be renewed, and it was rumored over the next few days that immense destruction had befallen Galveston.

Around 7 o'clock in the evening, the winds veered to the southeast and a five-foot tidal wave swept through the city. One moment the water was up to your knees, and the next, over your head. People quickly occupied the second and third stories of the crumbling houses. Consternation reigned almost to the point of madness as the winds were now estimated to be over 110 miles an hour, the anemometer having been blown away at eighty-four miles an hour. One survivor, Patrick Joyce,

Taking casualties to sea. (*Library of Congress*)

related a story of how he had been in a two-story cottage when it collapsed, sending fifty people into the surging waters. "I managed to find a raft of driftwood or wreckage and got on it, going with the tide, I know not where. I had not drifted far when struck by some wreckage and my niece was knocked out of my arms. I could not save her, and had to see her drown. I was carried on and on with the tide out into the Gulf. I drifted and swam all night, not knowing where I was going. At 3 A.M., I began to feel hard ground and knew I was on the mainland. I wandered around until I came to a house that was still standing and there a person gave me some clothes. I was in the water about 7 hours." Most of the city's 37,000 residents had similar experiences that night, and nearly a third of them did not live to tell, having been washed to death by waves and tides.

On the lighter side, one man—the superintendent of Galveston's public schools—drilled holes in the floors of his house, permitting the water to come in and anchor it rather than lift it from the outside. He said he drank a couple of beers and promptly fell asleep. He slept safe and sound, all night long, and even in his wildest of dreams could not have imagined what was taking place.

A British ship, the *Tauton*, was carried by the tide over nearby Pelican Island, "up the bay, along the shallows, far up where ocean vessels never went before and probably never will go again...until she was dashed head on to a bank 30 feet high...22 miles from deep water!"

It was two days before survivors realized the scope of their disaster. Identification of the dead was virtually impossible, and many were never found. Funeral pyres blazed in the city for days, and bodies had to be loaded on barges and tossed into the sea. The threat of pestilence was high, and it was difficult to get volunteers for this grisly task. There were literally thousands of human and animal bodies lying on the beach, decomposing in the late summer heat, and the stench was evident several miles inland. One refugee stated:

> I will not attempt to describe the horrors of it all; that is impossible. When I left Galveston, men armed with Winchester rifles were standing over burying squads and at the point of rifles compelling them to load the corpses on drays to be hauled to barges on which they were towed by tugs into the Gulf, and with weights attached, tossed into the sea.

As though that weren't morbid enough, many of those disposed of in this fashion were washed back to shore by the incoming tide.

Relief from a stunned nation was slow in arriving. One man stated, "The boats are gone, the railroads cannot be operated and the water is so high that people cannot walk out by way of the bridge across the bay, even should that bridge be standing." Several trains tried unsuccessfully to get to Galveston, but couldn't get within twelve miles of the city, as the prairie was covered with lumber and debris, pianos, trunks, and corpses. One train from Beaumont was blown into the bay by high winds, adding eighty-five to the death toll.

The Galveston seawall was completed in 1904, every building in the city being raised and fourteen million cubic yards of sand from the gulf being pumped under the raised structures. This project proved fruitful, for on the night of August 16, 1915, the city was swiped by a hurricane of equal strength, and the casualties numbered only eleven.

40.
CATASTROPHE AT LAKE OKEECHOBEE

FOR MORE THAN A WEEK, LATE IN THE SUMMER OF 1928, ALL EYES ON THE EAST COAST WERE **focused on a developing tropical hurricane that was making a name for itself in the South Atlantic. Actually, no effort was made to name individual hurricanes until the active season of 1950. Daily, headlines told of another Caribbean island that had been crushed, and residents along the Florida coast were wary, for two years earlier a very severe hurricane with winds at Miami Beach clocked at 132 miles an hour had caused untold damage throughout the Southeast and claimed 243 lives. The summer of 1928 unveiled the first commercial television, and had they been as common as they are now, everyone would have been on a first-name basis with this storm.**

During the week, Martinique and Guadeloupe had been leveled, along with most other islands in the Lesser Antilles. Darkly humorous stories were circulated as to the thriving business voodoo priests were doing selling charms against death, before the blow came. It was not until the United States territory of Puerto Rico was cut off from the world that this cyclone was taken with any degree of seriousness. Tales were told that erased the smiles from the faces of the very young and deepened the lines in older faces. John Tucker Battle, who had survived a similar hurricane just two years earlier, told what it was like:

> It's a ghostly sound like a winter wind makes in a chimney, so far and sustained. A small drum starts throbbing. The palm fronds whisper nervously and far out on the glassy yellow surface of the bay, a whitecap breaks like a puff of smoke. The sun is growing angry and red and it is becoming strangely dark. A soft rushing sound like a distant mountain torrent comes nearer and nearer. An angry puff of sulfur air kicks up spiral dust devils that dance up and down the street. Then with the rush and scream like thousands of express trains rushing through a metal tunnel, the hurricane strikes.
>
> Great strips of iron roofing are ripped off and hurled through the air. Giant palm trees roll along the beach like broom straws. The stout stone walls of the hotel groan and complain and the crash of clay tiling smashing against walls is like a shrapnel barrage. Long brown rollers sweep over a large area of what was less than a half hour ago a city. Fisher boats are smashed and tossed high into the center of town.

The original assessments of the damage in Puerto Rico were modest. News trickled in slowly, but in a few days Americans began to appreciate the horror of the event. More than one thousand people were killed outright, and another several hundred died from the ensuing starvation and pestilence of malaria, typhoid, and influenza. 400,000 people from that Caribbean island were left absolutely destitute of the basic necessities of life and relied on the Red Cross for food, clothing, shelter, and medical aid. Americans were quick to respond, but the problems were so overwhelming that they persisted many weeks and months after the hurricane departed. Elmer Elisworth, an American living in Puerto Rico, stated:

> As I stand on my home on the hilltop and look out over the hills and valleys, I do not see Puerto Rico, but a landscape that reminds me of the barren lands of Arizona or New Mexico. The country is blasted. Land under cultivation before the storm is now hard as concrete. There is no human life apparent, but at night here and there fires may be seen and where those fires are, homes had been....

Many deaths were caused by zinc roofs flying through the air like scythes. Sand was hurled from the beaches to homes a mile inland and came through the windows and blinded people's eyes. In the hilly country water was pouring down in streams through red sand which was so brilliant, it "looked like a river of blood." Oxen ran wild through the country, some of them still yoked, fighting and snorting in an effort to gain freedom. Lightning flashed constantly, and there was the continual roar of thunder and the crash of falling debris. The waves which swirled high in the air contained sand and mud and fell heavily on the shore. Then another one would leap and bound, and a reflection of fire in the sky could be seen in the ocean.

Meanwhile, Florida residents were preparing for the worst. On the night of the fifteenth, the hurricane was poised about 200 miles to the southeast of Miami and already the coast was buffeted with gale-force winds and battering tides. In its final dash to the coast, this storm packed winds of 135 miles an hour, which blew the Everglade flamingo clear across the Gulf of Mexico to Brownsville, Texas. The eye of the storm passed over land in the vicinity of West Palm Beach and turned extravagant mansions into rubble. The southern half of Florida is only a few feet above sea level, and monstrous waves routinely smashed windows, leveled homes, and tore away at the foundations of the sturdiest buildings. What was not immediately apparent, even as the storm turned and beat a destructive track up the East Coast, was that a mind-boggling catastrophe had occurred at Lake Okeechobee.

This body of water is the fourth-largest natural lake in America, and engineers had recently constructed a network of dikes to serve as a catch basin for water supply in Central Florida. It is not a very deep lake, stretching as it does in a circle, with a forty-mile diameter. The water level was usually maintained slightly above the land level so that it could be drained off by canals in normal times. It was said to be operating at this higher level when the hurricane hit, and the eighteen inches of rain which resulted was more than the new dikes could bear.

Suddenly everything just gave, and a ten-foot wall of water rushed into the towns on its southeastern side. It was a flood from which there was no escape. Six days later, when Red Cross volunteers finally penetrated the town of Pelican Bay, they discovered that each of the 450 residents had perished. In other places, where there were survivors, grim tales were to be heard. People who had witnessed the drowning of other family members and the destruction of their homes had to wade or swim through five feet of water for six miles before the waiting ambulances shuttled them to safety. The Everglades, which had al-

A Haitian village lay in ruins following the passage of an intense hurricane on October 13, 1928. (*LIL*)

ready been ravaged by wind, were then inundated by the waters of Lake Okeechobee. There were even reports that alligators infested the canals and were eating the bodies of the dead.

Buzzards circled over Lake Okeechobee, betraying the presence of the many who died there. The bodies of drowned snakes and turtles floated in the stagnant waste, and the area was quarantined. One account described the lake as a "muddy brownish fluid too heavy and smelling of corpses of man and beast on its breeze tossed surface to be called water." The individual stories are most tragic. Clare Schlecter, a wan girl of fourteen, was sitting listlessly in an effort to realize that she was alone in the world, six other members of her family having drowned. Max Yarborough, a boy of twelve, was having his head bandaged. He had watched his mother, brother, sister, and aunt sink in the swirling waters while he was pinned beneath the broken timbers of their home. Perhaps the most pitiful figure of all was a W.W. Britt, sixty-two, who kept calling for his wife of seventy. Arm and arm, they had clung to floating timbers when their home was wrecked by wind and wave. Only 200 yards away from it, timbers slashed against them, and Mrs. Britt was carried off in the swirling mass of debris, out of her husband's reach.

In all, more than 1,650 lives were blotted out in this community, and the storm totals for the entire swath through the South Atlantic approached 5,000. Americans living along the East Coast have not witnessed a hurricane of this stripe for more than 20 years, and most have never seen one. Just as the city of Galveston built a seawall and engineers took another look at Lake Okeechobee, many more have settled into the coastal waters from Maine to Texas, and its unthinkable to imagine the toll should a hurricane slam into a populated area now.

41. THE HURRICANE THAT LOST ITS WAY SEPTEMBER 21, 1938

THE SUMMER RESIDENTS IN NEW ENGLAND WHO THOUGHT HURRICANES WERE SOMETHING **that happened to Florida at a time of year when practically no one was there were in for quite a surprise when dawn broke along the fashionable resorts on the twenty-first of September, 1938. The day after what was undoubtedly one of New England's outstanding meteorological events, they found themselves digging out of the splintered wreckage of their homes and cottages wondering what went wrong. For many years, this area of the country had been virtually immune to tropical storms, and it was not within the memory of the oldest inhabitants of a more spectacular storm.**

This hurricane went virtually unnoticed after sideswiping a few Caribbean islands over the weekend. The Northeast had endured a particularly soggy spell with rain falling continuously for five days in spite of occasional appearances of a late summer sun. In Philadelphia, the day before the storm, 3.64 inches of rain were measured in a tropical cloudburst. Long Island was already flooded prior to the hurricane, which on the afternoon of the twentieth was making slow progress northward about 600 miles off the Georgia coast. By the morning of the twenty-first, it had dashed forward nearly 650 miles to a point well off the North Carolina coast. In the forecasts for that day, there was no mention of rain as modern electronic surveillance methods were unknown at that time. The storm accelerated to more than fifty miles an hour in forward speed and smashed into Long Island at 3:30 in the afternoon, its first U.S. landfall. Striking with practically no warning and with unprecedented strength, that storm blotted out more lives and caused more property damage than any other event to date. The hurricane made the entire trip, from the southern coast of Long Island to the Canadian border in less than five hours.

Along the New Jersey coast, storm waves were estimated at greater than thirty feet high, carrying miles of boardwalk several blocks inland. On Long Island, the Suffolk County town of Westhampton was particularly hard hit, as one resident recalled:

> I thought it was an earthquake. The whole house started to tremble...and we began to be afraid that something was going to happen. I was never so frightened in my life. The wind was howling and water was up to the floor in our living room. I bundled the baby in a blanket and we set out together... water swirled about our hips while planks, branches and all sorts of things were hurled through the air. It was only by the grace of God we were not killed.

With almost mathematical precision, this tree was quartered during the New England hurricane of 1938. (*New England Historical Events Association*)

This family struggled inland a half a mile to escape the infringing ocean, and they were all saved. They looked back once to see their large rambling home on the beach collapse with a roar. Nineteen others in this small resort community lost their lives. In Patchogue, the barometer set a new all-time low of 27.95 inches. Even New York City was not spared from this blast, where the storm attained its mightiest fury between 4 and 5:30 P.M. One man, who set out from the city for Montauk, was not aware of anything other than an unusually heavy rainfall at 3 P.M. Halfway out the island, torrents of rain were falling, and every now and then a tree would fall across the road and he would duck out of his car, quickly survey the situation, and then gingerly make his way around the obstruction. Every now and then a car would slide into a ditch. He told of one man in Southold who "vented his wrath at a fallen horse chestnut across his lawn. The wind blew every chestnut off his tree and through his windows, riddling them like machine gun bullets." This intrepid traveler, though unhurt, was finally stopped by the incoming ocean just a few miles out of Southold. He concluded, "There's a lot of work ahead for some people."

It was in Providence, Rhode Island, however that the storm was to take its greatest toll. Winds everywhere exceeded 125 miles an hour in New England, and a 186-mile-an-hour gust was measured at Harvard University's Blue Hill Observatory. That afternoon, 200 people were swept out to sea and drowned, as a seven-foot tidal wave struck the dune at Watch Hill, Rhode Island, where they had scurried, for the time being, to safety. The same storm surge inundated the city of Providence, sending afternoon theatergoers and shoppers to the second floors. Many others had to swim to safety, so sudden was the gush of water. The second-floor lobbies seemed comparatively safe until the waters of Narragansett Bay flooded downtown Providence to a record thirteen feet, 8.5 inches high. Thousands of people were marooned until the water subsided the following morning. The death toll in Rhode Island of 380 was not caused so much by the high winds of the hurricane as by the suddenness of the storm. By now, it was cutting through New England at a speed in excess of sixty miles an hour, and that

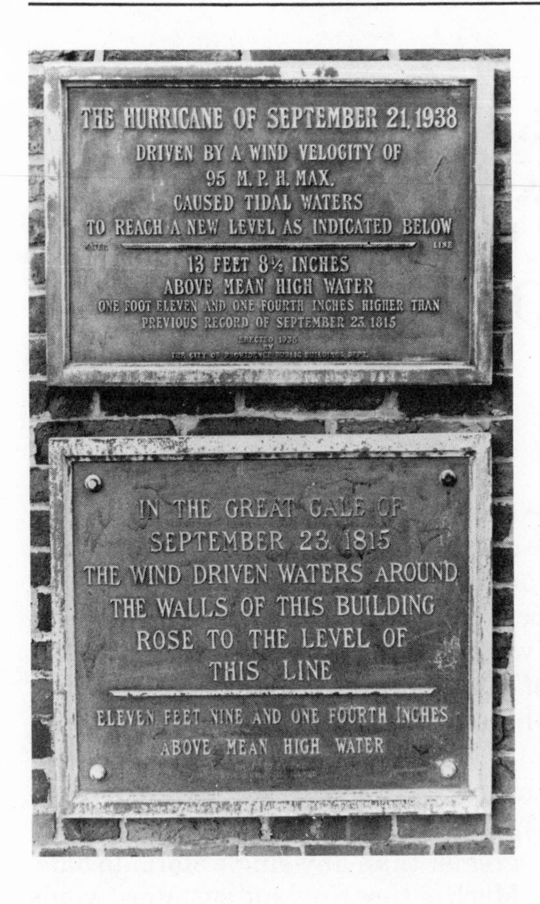

Plaques on a building in Providence, Rhode Island. (*Ron DerMarderosian*)

caused the waters to rise to record levels in less than one hour.

Even *Old Ironsides*, the ancient frigate, was torn from its moorings in Boston Harbor and damaged, so great was the onrush of the Hurricane of 1938. By the time the vacationers from the islands along the coast had told all the tall tales of whose misery was worse than another's, they realized that they were the lucky ones, since most of the damage occurred inland where the full-time residents were used to bad weather as one of the facts of life in New England.

42.
FEAR STRIKES OUT – THE INCREDIBLE HURRICANE SEASONS OF 1954-55

When I was a youngster growing up in Long Island, I thought that at least two or three hurricanes were expected to pass overhead every fall. The rains would turn the suburban streets into raging brown rivers and topple trees routinely. Home from school, I watched the havoc from the living room window and developed the habit of nibbling on the window sill every time another limb came crashing down, and I watched it get caught up in the torrent that was my street and disappear into who knows where. It was much later that I appreciated the anomaly of so many storms.

Actually, baseball intrigued me more at that time, for it was the golden era for someone living so close to New York, which sported no fewer than three teams, with stars such as Willie Mays, Mickey Mantle, and Duke Snider. On one of the few autumn days when it wasn't raining, I saw Don Larson pitch a perfect game in the World Series. One of the most vivid memories of childhood I have is the day after hurricane Connie passed out to sea; I walked to the park where Roslyn's historic Mill Pond was and saw a huge gaping hole in the concrete barriers of the pond, washed away by the heavy rains. The picturesque mill dam that had been the home for hundreds of ducks and geese was no more. Instead, they waded disconsolately about in the black muddy pond bottom, unable to grasp the significance of their disaster.

In the late summer and fall of 1954, the East Coast had witnessed three full-blown hurricanes, bearing the names Carol, Edna, and Hazel. Carol was the first, and it bisected Long Island and New England on the thirty-first of August, causing property losses greater than any single storm to date. Much of New England sustained winds ranging from 100-135 miles an hour, and I secretly envied my sister's good fortune to bear her name. Less than two weeks later, hurricane Edna came along following a similar path, but further east, smashing into New England's capes. Finally, in the middle of October, Hazel smashed ashore over the Carolinas and cut a deadly swath through central Pennsylvania. Washington, D.C., recorded its highest wind velocity of ninety-eight miles an hour in this storm, as did New York City with a 113-mile-an-hour gust. Most hurricanes become greatly diminished from the friction of passing over land, but this storm seemed to delight in adding power as she moved northward into western New York state and on into Canada. She spent much of her fury in the Finger Lakes district and still had enough punch to kill thirty-six more, north of the U.S. border.

Storm-driven waves batter the coast of Provincetown, Massachusetts, prior to the passage of Hurricane *Carol* in August 1954. (*New York Public Library*)

It was no wonder that in August of 1955 the East Coast was jittery, as a giant of a storm, Connie, packing winds of 135 miles an hour, was poised only 200 miles off the South Carolina coast, spinning like a mad top. The entire coast from Jacksonville, Florida, to Provincetown, Massachusetts, was on alert as gales blew nearly 250 miles from the center in every direction. It was early in the era of hurricane study that Captain Edward L. Foster flew his plane into the eye of this storm and described what he saw:

> In the eye, you would think you were sitting in the middle of a huge amphitheater. All around us in a huge circle were bands of white clouds. Below was a deck of stratocumulus clouds and above was the bright blue sky...we flew up 10,000 feet and the walls of the amphitheater still rose above us. It was hot in the center too, 86°, which is hot so high in the air, and the humidity was at the saturation point.

It was calculated that Connie was expending the energy of a few million atomic bombs—enough energy squandered to keep America lighted for three million years or enough in one minute to keep the country in electricity for fifty years. Long before she came ashore, Connie was already spawning tornadoes inland over the Carolinas and chasing bathers from the beaches as far north as Long Island. Giant breakers pummeled the shoreline three days in advance of her arrival, as her unwillingness to commit her striking force kept the entire Atlantic Seaboard in a frenzy for several long days. Diane, her younger brasher sister, nearly 1,000 miles southeast of the U.S. mainland, moved very little and exercised a restraining influence on Connie. While Diane sulked in the South Atlantic, weathermen saw a perfect example of the "Fujiwara effect," which is when the circulation of two storms collide and one storm seems to rotate around the other. It was this influence that held Connie back and drew Diane directly north towards the East Coast.

An anemic image of her former self, Connie finally grazed North Carolina and came inland through the waters of the Chesapeake Bay on the thirteenth of August, 1955. Her windy arms buffeted the Mid-Atlantic and New England with winds averaging only seventy-five miles an hour, but her unanticipated punch was the record deluges throughout the Northeast, including twelve inches on western Long Island, wash-

A helicopter takes a man to safety while another awaits his turn at a top window at Scranton, Pennsylvania. (*ILN*)

At Uxbridge, Massachusetts, a motorist (center, right) is pulled through the flood waters after he had been trapped in his car. (*ILN*)

Transformed into an island sea by the rain-swollen waters of the Potomac, a scene in East Potomac Park at Hains Point, Washington, D.C. (*ILN*)

Hurricane *Edna*, the third storm of the season, was responsible for collapsing this structure like a house of cards on September 12, 1954. (*National Archives*)

ing away most August rainfall records. By nightfall, patches of blue sky shone through, although the gloomy gray apron of Connie would obscure them with one final lashing of her eastern fringe.

The summer of '55 was the hottest known in so many places up and down the East Coast that temperature records were shattered almost daily. The usual concomitant of heat—namely drought—had been relieved from Connie's rains, but this rainfall saturated the ground and filled the streams, thus setting the stage for Diane. She had now taken a left turn towards the United States mainland and neither her strength of 115 miles an hour nor her course was very comforting.

Diane finally struck the coast early in the day of the seventeenth, just five days after Connie had grazed an area 200 miles north, in Fort Macon. She plunged northwestward, her winds diminishing, and the northeastern United States heaved a sigh of relief. Diane was counted out and, for one of the few times this memorable summer, weather failed to make front page news. It should have. The weather forecast in the New York *Times* in the August eighteenth edition read, "Fair and hot, today and tomorrow." Somewhere along the Appalachian Trail in Virginia, Diane changed course and made a beeline for southeastern Pennsylvania, New Jersey, and southern New England. Her unadvertised kick was far worse than any wind.

Diane sucked in vast quantities of moisture from the Atlantic and dumped water down by the ton in advance of her now leisurely pace. The runoff from the rains was at a maximum, and everything from large rivers to minor tributaries began to swell from thirty-six-hour rains which dumped fourteen more inches in Hartford and fifteen inches in Boston. The rivers near Hartford rose 19.5 feet, and "you could hear boulders crashing down the hillside onto the road." In nearby Putnam, the Metal Sellers Corporation caught fire and 200 tons of exploding magnesium were carried on the flood waters out of the building and into the business district. It was a night of terror for the residents from dusk to dawn, for other fires were set and showers of fiercely blazing metal, some fountains 250 feet high, brightened the sky. Firemen stood by helplessly, unable to penetrate the raging flood waters near the building.

On the very same night, the eroded hilly country in northeastern Pennsylvania was especially hard hit, as the banks of the Delaware, Susquehanna, Schuylkill, and Lehigh could not begin to cope with the rush of water from the hills. Dams burst, but the most agonizing news came from a summer camp in the Poconos. It was the story of a flood which burst upon Camp Davis about three miles northwest of Analomink, sending the Brodhead Creek up twenty-five to thirty feet in just fifteen minutes. As the children and counselors struggled to reach higher ground, their bungalows were swept away, and they watched from the second floor of a nearby building. One of the staff of the camp, a Mrs. Nancy Johnson, described the scene:

> We were terrorized but couldn't do anything but watch the water coming up toward us. It kept getting higher and higher and when we reached the attic, we felt the house give a shudder and the whole house collapsed. It just fell apart and the 40 of us went tumbling into a jumble of water, boards and screams.
>
> It was dark but I could hear the children screaming for help. I wanted to help them but it was all I could do to keep myself on an even keel. It must be awful to drown. I went down, down, down and I guess I kept waving my arms to get back to the surface.

Mrs. Johnson was knocked out by debris, and when she came to, she found herself floating on her back. She quickly grabbed a board, which she shared with one of the children.

> Beth and I held on to each other and cried every time we heard a child in the distance crying for help. Some of them screamed hysterically and I could not tell if any of the voices belonged to my own children.

In all, thirty-one lives were claimed—mostly children—and many Pocono residents, their ears still echoing with cries of "help me" or "save me," walked the streets red-eyed and dazed. In all, at least 184 people were accounted dead as a result of this storm and its devastating floods.

In the following days the heat resumed, and on the twenty-first, just two days after the departure of Diane, a severe squall line hit the East with sixty-mile-an-hour winds, knocking over trees and exposing live wires once again. Not content with hurricanes, drenching rains, devastating floods, record heat, and surprise line squalls, the elements of August 1955 conspired to set more records of all sorts in the Northeast than any other known.

43.
MOVING DAY—HURRICANE *CAMILLE* AUGUST 17-18, 1969

THE SUMMER OF 1969 MARKED THE WOODSTOCK EXPERIENCE, WHEN HUNDREDS OF THOUsands attended the greatest outdoor rock concert of all time, to see the likes of Joan Baez, Ravi Shanker, Jimi Hendrix, and the Jefferson Airplane perform. For three days, they camped out in the mud of an Upstate New York farm, and it was on August seventeenth of that year that they began their grand exodus, for their lives had been changed and it was time to move along. On the southern Gulf Coast of America, another mass exodus of sorts was taking place involving a similar number of people. Hurricane Camille, rated the second strongest to prowl the Atlantic, was making a move at the southern shores of Alabama, Mississippi, and Louisiana with her 190-mile-an-hour winds. It was an event destined to change the lives of hundreds of thousands of Gulf Coast residents, and the time to evacuate was at hand.

One man, newly transplanted from the Midwest and unfamiliar with the wrath of a hurricane, stopped in a local hardware store in Gulfport, Mississippi, and asked what could be done to prepare for the storm. The reply, although somewhat tongue in cheek, was a typical one practiced by many: "Get yourself a bottle of good whiskey and set back and relax!" Hundreds, perhaps thousands of people, staggered through Camille in drunkenness, but many found that whiskey does not always work. There was quite an eyewitness story printed in the *National Geographic Magazine* several years back about some friends who decided to have a "hurricane party" at a waterfront apartment. About two dozen revelers showed up to ride out the storm, but only three of them survived to weather a hangover. The rest were swept out into the gulf by twenty-foot storm tides.

Hurricane Camille came ashore in the Mississippi Delta region, a narrow strip of land which protrudes about 50 miles into the gulf. The town of Buras, the first to experience Camille's power, was soon to be wiped off the map. Jutting well out into the gulf, it caught the entire thrust of Camille's 190-mile-an-hour winds and storm surge, and in the words of Luke Petrovich, the public safety commissioner, "It's gone—not destroyed—it's gone. There is no more Buras, it's all gone, flattened, nothing, just nothing." When Camille hit the mainland, the story was the same. Gulfport, Biloxi, Pass Christian, and Bay St. Louis were all demolished, raked by tornadoes and unprecedented winds. The storm began whipping New Orleans at dusk, which came early under scudding steel-gray clouds, and the world's largest bridge, the twenty-six-mile causeway that spans Lake

Flood victims line up for food and pure water. Said one relief worker: "It ain't going to hurt the government to feed and clothe them that needs it." (*New York Public Library*)

Ponchartrain and links New Orleans with the outside world, was under water by dark. Winds lashing the bridge approached sixty miles an hour at that time, and the lake "was in an uproar of high wave and froth." One man recalled, after his house had crumbled from the onslaught, that "3 or 4 times as we went down, my wife would come up and say 'go on and leave me.' Well, I wasn't going to do that. I sure didn't have any place to go without her."

The waters rose with disastrous force and speed, and fortunately for most they had retreated to Hattiesburg, about 100 miles north of New Orleans. Nearly every house, every tree, even the oaks that had survived a hundred years of gulf storms, gave in to this one. Pine trees 100 miles inland were all snapped in half. Water moccasins and cottonmouths by the thousands were washed inland and slithered about on suburban lawns, in the salt and squalor of

A young man considers his loss after Hurricane *Camille* (1969). (*New York Public Library*)

the hot delta sun. One survivor testified that the wind was blowing rocks and snatched three large ocean liners from their moorings and set them down three blocks inland, nosed together like rowboats. The Gulf Palms Hotel in Pass Christian, a sprawling U-shaped vacation haven, was left without a brick.

When it was all over, people slowly crept back into their communities to see if the worst had been realized. Many stood in front of their houses looking embarrassed as disaster victims often do, suffering from the discomfort of having been caught in bad luck. One man returned to inspect his house and said, "It's gone. There's nothing left...I just turned away and left it for anyone who wanted it." Poisonous snakes roamed the towns, making them almost as dangerous as when the hurricane struck.

Their faces were all the same: grim, gaunt, and tight-jawed. Men and women came in dozens of buses to inspect the damage, and when their eyes confirmed what they had been told, the buses turned around without stopping and headed back for the Red Cross shelters in Hattiesburg. The National Guard was called in to clean up and keep order. The fumes from the multiple gas leaks in the many towns prompted one guardsman to remark, "One spark and we won't have to worry about cleaning up Buras anymore." He wasn't exaggerating. Another man in his seventies inquired, "Anyone going to Belle Chase? I got to get me a ride to Belle Chase to the shelter. I ain't got a home no more... couldn't find a trace of it. It's gone, out there in the Gulf some place. Got to get me to Belle Chase for a drink and a sack to sleep in." Carrying his possessions—two pair of pants and a box of Alka-Seltzer—he trudged off looking for his ride.

Hurricane Camille probably did more for integration than any other event in the history of the state of Mississippi. Banded by calamity, blacks and whites were living and eating together in shelters and working side by side as though the racial issue no longer mattered. One black man, whose hands had been blown off by a gas explosion, remarked that he "ain't able to work...ain't able to help. Gonna sit here and cry, I'm afraid...." And he did cry, and was ashamed to show that he could not stop. There were many days spent in horror, numbing confusion, and incredulity along the Gulf Coast in 1969, and many were slow to respond to the wild scenes of devastation that surrounded them.

Meanwhile, hurricane Camille took a run up the Mississippi River and turned east for Virginia. While much of her wind dissipated on the nineteenth, rains began to fall heavily upon the Appalachian Mountains in central Virginia. Twenty-seven inches fell overnight, despite forecasts of "occasional showers," and severe flash floods resulted. The rush of water from the Blue Ridge Mountains was so fantastic that people in its path were left with little or no warning. One woman was awakened by the sudden rush of water and looked to her window to see a Black Angus cow staring back at her, apparently dead but floating right side up in the awesome flood. The agricultural area between Charlottesville and Lynchburg, Nelson County, was inundated when the James River suddenly rose fifteen feet in twenty minutes.

Camille's backlash, in the state of Virginia alone, accounted for 109 of the hundreds of victims claimed by this storm. Despite traveling overland for hundreds of miles, Camille retained the punch that ranks her among the most powerful storms of any sort to cross the United States.

44.
HURRICANE SAFETY RULES

THE NATIONAL HURRICANE CENTER IN MIAMI HAS OVERALL RESPONSIBILITY FOR THE monitoring of hurricanes in the Atlantic, Caribbean, and Gulf of Mexico. It issues watches and warnings for affected coastal areas while the responsibility subsequently passes to the Washington office when the storm moves north of Cape Hatteras and to the Boston office when it moves north of Long Island.

A ***hurricane watch*** is issued for areas where a hurricane is a real possibility, but not imminent. People within the watch area should take preliminary precautions and be prepared to act quickly should a warning be issued. A ***hurricane warning*** indicates that hurricane conditions are expected within twenty-four hours and immediate actions, such as evacuation of low-lying areas, should begin.

1. Enter each hurricane season prepared: recheck your supply of boards, tools, nonperishable foods, and other equipment you would need should a hurricane strike your community.
2. When your area receives a hurricane warning, plan your time before the storm arrives and avoid the last-minute hurry that might leave you marooned or unprepared.
3. Leave low-lying areas that might be swept by high tides or storm waves.
4. Leave mobile homes for more substantial structures.
5. Moor your boat securely before the storm arrives.
6. Board up windows or protect them with storm shutters or tape.
7. Secure outdoor objects that might be blown away or become deadly missiles in high wind.
8. Store drinking water in clean bathtubs or jugs. Your town's water supply may become contaminated by flooding.
9. Check your battery-powered equipment: your radio, flashlight, and cooking equipment may be essential if power is interrupted.
10. Keep your car fueled. Service stations may be inoperable for several days after the storm strikes due to an interruption of power.
11. Stay at home if it is sturdy and on high ground. If not, go to designated shelter areas.
12. Remain indoors during a hurricane; travel is extremely dangerous.
13. Beware the eye of the hurricane. If the calm center passes directly overhead, there will be a lull lasting a few minutes to a half hour or more. Stay in a safe place unless emergency repairs are necessary. On the other side of the eye, winds rise rapidly to hurricane force, from the opposite direction.
14. When the hurricane has passed,

seek necessary medical care at Red Cross disaster stations or hospitals.
15. Stay out of disaster areas unless you are qualified to help.
16. Drive carefully along debris-filled streets. Roads may be undermined and collapse under the weight of a car.
17. Avoid loose or dangling wires and report them immediately to your power company.
18. Prevent fires; lowered water pressure may make fire fighting difficult.
19. Check refrigerated food for spoilage if power has been off during the storm.
20. Remember that hurricanes traveling inland can cause severe flooding. Stay away from river banks and streams.

Courtesy of the U.S. Department of Commerce, National Oceanic and Atmospheric Administration, Public Affairs Office

VIII. SOME OUTSTANDING WEATHER EVENTS

45.
SAXBY'S GALE

NOWHERE IN THE HISTORY OF METEOROLOGICAL PREDICTION CAN ONE FIND THE EQUAL OF A prognostication issued by Lieutenant S.M. Saxby in late November of 1868 in the London press. Based upon astronomical events which were to occur eleven months hence, in October of 1869, he forecast that a tempest of unusual strength and high tides would strike in the early morning hours of October fifth, somewhere in the world. He based his prophesy on a lunar coincidence that would place the earth's satellite precisely over the equator, during its closest approach to the earth. He reasoned that the combined gravitational effect of the sun and moon at that time would bring a tide of record proportion, and his prediction was taken with some seriousness in scientific and popular circles, as he had had enormous success with this sort of thing in the past.

This dire prediction was subsequently taken up by a certain Frederick Allison of Halifax, Nova Scotia, and further refined. Allison pinpointed the event to the Halifax area on the Atlantic coast of Nova Scotia for exactly 7 A.M. on the morning of the fifth, and the fulfillment of his forecast comprises perhaps the best long-range forecasting job, before or since. The storm, perhaps of hurricane origin, can be traced only as far south as the Outer Banks of North Carolina, where rain commenced overnight, Sunday, October second. Shortly before daybreak, the first waves of the storm were felt in Washington, D.C., and pushed up the coast to Baltimore and Philadelphia shortly afterward.

Perhaps Americans were not so well attuned to the Saxby or Allison predictions, and those who had heard of them undoubtedly made no connection of the current rains, for so much time had passed. Besides, there was no advance warning that a full-fledged hurricane was churning northward off the coast, for reports of weather at sea were virtually unheard of, considering the state of the art in the middle nineteenth century. By midday on the third, the rains increased and the wind became violent along coastal areas, betraying the presence of a tropical disturbance. By evening, the streets of Baltimore lay flooded, and residents feared a repeat of the recent July floods in that city. Alarm bells were set clanging as backwater from the city's sewers gushed into the streets. In Philadelphia, the incessant, driving rains produced a major flood of the Schuylkill, through the main arteries of town. Lumber and coal were carried away from the wharves, downstream, as water reached the second storey of the Schuylkill Navigation stores. About twenty miles upriver, in Norristown, residents noted a seventeen-foot rise in the Schuylkill, and most of the downtown area had to be evacuated.

Drought had become the catchword in New York City, as their reservoirs had dropped off substantially in September, the water in Central Park having dipped nearly twelve feet beneath the top of the dam. After the storm's passage, the *Times* noted, "It is consolation to know that it (the storm) has been sufficient to avert a state of affairs...that was threatening to put our people to the most serious inconvenience." The temporary inconvenience involved the loss of power, communications, and transportation, as most railroads were completely submerged or washed away.

Further north, Saxby's Gale inflicted far more damage. All six New England states recorded more than six inches of rain as did Virginia, parts of Pennsylvania and Delaware, and New York. The gale precipitated the greatest New England floods known in the nineteenth century. The enormous power of this storm was magnified by a slow, eastward-moving weather front that acted as a wedge to the moisture and lifted it high above the condensation levels of the atmosphere to produce some astonishing rainfall amounts. Canton, Connecticut, on a hillside location which normally afforded healthy rainfall amounts, collected more than twelve inches of rain in a twenty-four-hour period, a feat which went unchallenged until the passage of hurricane Diane in 1955.

A rather astonishing story came out of Hoosick Falls, New York which told of an ill-fated passenger train that had pulled out of a station in Lansingburg and collided into a freight train after only a few miles of their journey. None of the passengers were seriously injured, and they exchanged seats when the next train came through, abandoning the wreckage of the mother train. A couple hours later, upon reaching Eagle's Bridge, which had been undermined by the steady rains, the trainman, H.J. Clark, pulled to a halt. He radioed the other side of the bridge, where his brother Robert was a conductor on yet another train, and he proposed that all fifteen passengers and crew take conveyance on his train to Hoosick Falls. This involved walking over the weakened bridge, and all aboard assented to the venture. Engineer Clark slowly backed his train to within a short distance of the falls when the shouts of "stop—back up" were heard. He threw the engine into reverse, but it was too late. The engine settled into the sinking track and plunged fifty feet down the embankment into the boiling waters of the Hoosic River.

It is extraordinary that anyone made it out alive, for the train actually somersaulted as it tumbled, and many of the dead had received serious head injuries. Several more drowned in the surging waters of the Hoosic, while both Clark brothers survived the fall, as did one other passenger. Lying in the wreckage, they managed to extricate themselves from the track that had pinned them and crawl to the safety of the steep embankment. Engineer Clark recalls that he would have passed out from injuries and shock had he not been revived by the steady rain on his face. In the same town on the same day, there had also been a mining disaster caused by high waters slashing through a coal shaft, and similar stories of daring and rescue came from that sight.

Much of Eastport, Maine, was smashed into atoms as hurricane-force winds and tidal waves tore away at the town's wharves, fish houses, and stores. Twenty-seven vessels were driven ashore at Rumney's Bay. All the houses and barns in nearby Lubec, Pembroke, and Percy were obliterated as the press reported that "this tornado was worse for Eastport than the great fire." Outside of Lewiston, a normally tranquil Swift River, a winding mountain stream, rose thirty-six feet in only twelve hours, nearly one-third of that occurring in a mere half hour. Most people barely escaped with their lives, bearing children in their arms on a flight to the mountains. So overwhelming were the angry waters that several families were literally swallowed up in minutes, without notice, as the Swift River became a sea of death and destruction engulfing the earth. One elderly couple was trying to save their harvest of corn and pumpkins when some lumbermen nearby urged them to flee for their lives. Stubbornly, they persisted even though the water was over their waists and rising rapidly. One man swam about sixty rods out to save the older farmer and pulled him to safety. Rescue of the woman was impossible, and she spent the night in a tall tree praying. All night, she related, the water lapped at her face before receding the next morning. The lumbermen came back to rescue her after the flood waters had receded to about three feet.

Throughout New England, barely a bridge was left intact, and so many dams and dikes were washed away that hardly a windmill or waterwheel was left turning. Many trains that had left on the day of the fourth either met a tragic end or were floodbound for the next three days. The storm raged all night long and into the following day in New Brunswick and Nova Scotia, and a tidal bore in the Bay of Fundy produced the highest tides ever known in that region, precisely at the hour predicted by Saxby and Allison. To be forewarned is not always to be forearmed.

46.
A TERRIFIC AUTUMN GALE

IN THE EARLY YEARS OF THE 1870S, THE WEATHER BUREAU WAS STILL KNOWN AS THE **Signal Corps, and the Coast Guard would not be formed for another forty-five years. This situation left those in charge of monitoring the weather and its consequences with archaic tools, with the benefit of a century's retrospect. The Signal Corps was organized into a group of 147 observation stations that depended upon the nation's fragile network of telegraphic devices which were as often as not put out of commission when they were needed the most, at the height of a particularly devastating storm. The flow of information was often impeded for several days following the passage of a particularly violent storm, and the science of meteorology, such as it was late in the nineteenth century,** was rendered silent. Nevertheless, the Signal Corps was proud of their accomplishments and boasted an eighty-four percent accuracy rate, with the figure of ninety percent considered attainable. A recent visit to a local Weather Service office disclosed the very same percentages and hopes. Despite the giant advances in technology, an accuracy rate of eighty-five to ninety percent over a forecast period of twenty-four hours is still considered pretty good. Much of this book is composed of the outstanding ten to fifteen percent of the weather that surprises even those whose livelihoods depend upon the success of their forecasting abilities.

In the same era, the U.S. Life Saving Service came into existence as a forerunner to the Coast Guard. Built seven miles apart, each beach station had boats, beach carts, and horses at its disposal during the autumn and winter seasons, a crew of seven or eight men, and a keeper. When a ship in distress was spotted, men nearest the station hauled their oar-powered boats to the beach, launched them into the thundering surf, and rowed to the stricken vessel under the worst of conditions to take as many survivors as could be had, through the heavy surf to the shore. On a routine night, a beach patrol was organized with one member of one life-saving team walking three and one-half miles to the halfway meeting point, where a member from the team seven miles down the beach would greet him and exchange signatures indicating that the two had fulfilled one of the requirements of their jobs. For seven lonely miles, round trip, these men braved the elements, scanning the sea for ships in distress. Should one be spotted, the remainder of the crew was alerted, and boats would occasionally have to be hauled three miles along the beach before launching towards another rescue. Needless to say, many of these men perished in the line of duty.

Both the Signal Corps and the U.S.

Life Saving Service were pressed into extraordinary action when a storm of tropical origin lashed the East Coast in October of 1878. As early as the twentieth of the month, storm warnings flew in Key West as an approaching cyclone from the South spread ominous clouds up the Florida peninsula and rough seas in its advance. In the Southland, in the vicinity of New Orleans, a Yellow Fever epidemic was rampant from early July until the latter days of October, and had already claimed 4,000 lives and prostrated another 12,000 people. The Gulf Coast was anxiously awaiting the season's first frost, to put an end to this still mysterious disease. In the wake of the hurricane that passed to their east, enough Canadian air penetrated the bayous, and their wishes were granted.

Residents along the East Coast did not fare so well. By the twenty-first, the storm had moved to a position just southeast of Key West, and on the following morning, it was just offshore, east of the southern Georgia coastline and intensifying rapidly. By midnight, on the twenty-second, the tropical storm was in the vicinity of Cape Hatteras, and much of the Outer Banks were submerged under the thundering waters of the ocean. Smithville, North Carolina, picked up more than 4½ inches of rain, and this amount was common in the storm's northward drift. By the morning hours of the twenty-third, residents of Cape May, New Jersey, were astonished as a full-force gale of eighty-four miles an hour was in progress, with tides covering the inland meadows and escape to the mainland all but cut off.

Philadelphia and the surrounding countryside of Pennsylvania were especially hard hit, as winds rose to hurricane force for six hours on the fateful morning of October twenty-third. As the storm attained greater violence, hundreds of buildings were literally blown down in the Quaker City, and flooding of the Schuylkill ensued. Nearly fifty churches lost their spires, several of them towering nearly 180 feet in the air, only to come crashing down on neighboring dwellings. Public squares, schools, factories, and warehouses alike were reduced to ruins on that morning of unprecedented wind, and it was estimated that more than 500 buildings and more than a dozen major bridge-spans were demolished in the autumn gale of 1878. The winds produced by this storm in Philadelphia that day still rank number one for sustained wind speed.

Railroad depots in the City of Brotherly Love were uniformly demolished. The Pennsylvania Railroad Station across the river in West Philadelphia was leveled; the 12th- and 16th-Station passenger depots entirely demolished; the Philadelphia-Wilmington & Baltimore depot reduced to rubble. Further upriver in the usually flood-prone districts of Wilkes-Barre/Scranton, a tornado striking at 10:30 A.M. lifted up a three-hundred-by-ninety-foot building owned by the Lackawanna Iron & Coal Company, bringing the entire place down "with a terrible crash at a single swoop." In nearby Pottstown, the newly erected Madison Bridge which spanned the Schuylkill was washed away by the high floods, and damage in that small hamlet alone was estimated at over $100,000.

Along the coast, life saving squads were kept busy for several days, pulling out one survivor after another from the life-engulfing waves which swooped in to shore at an estimated twenty-five to thirty feet high. Many an effort to scale the towering waves met with failure, and these valiant men could only watch in frustration the fate of seamen on craft which had split up as a result of the monstrous wind and wave. There were other stories, too numerous to mention, of astounding rescues at sea, and many an abandoned ship sailed along for hundreds of miles on a ghost journey in the high seas. The steamer *Express*, weighing 350 tons, was swept out to sea near the mouth of the Chesapeake and swept out to sea before breaking up without a sole survivor. One hundred and two years later, on this date, the S.S. *Poet* disappeared in these waters with the same number of sailors aboard—twenty-nine—never to be heard of or seen again.

It was a storm of unusual violence for this season, for the end of the hurricane season normally comes in the first or second week of October. That the wind records still stand in many eastern localities from this storm is a lasting tribute to the mighty forces at work in the South Atlantic in the year 1878. The storm eventually blew the infested mosquitoes clear out into the Gulf of Mexico, and the monumental Yellow Fever epidemic in New Orleans subsided for the winter. Blessings sometimes come in big packages.

47.
THE LOSS OF THE S.S. *PORTLAND*

SELDOM IN AMERICAN METEOROLOGICAL HISTORY HAVE SO MANY PEOPLE DISAPPEARED with such mystery surrounding their fate as in the Great Atlantic Gale of 1898. On the twenty-sixth of November of that year, a storm of considerable strength was developing along the Mid-Atlantic coast, moving slowly northward on a fateful meeting with the S.S. *Portland* that had just embarked into the frigid stormy seas. Not one of the 191 passengers aboard the ship was to survive the storm, and nearly 250 others perished at sea overnight in the gale winds spawned by this incredible autumn tempest. Along the coast, windswept rains lashed the residents along the East Coast, while west of the storm track, heavy snows prevailed for the second time in a week, culminating in a twenty-seven-inch fall in New London, Connecticut.

In the troublesome period between November twenty-second and December fifth, four storms pounded the East Coast with record-breaking cold and snow to mark their presence. After a blizzard swept the Great Lakes region on the twenty-second of the month and reformed into a moderate snowmaker along the East Coast, dropping from six to twelve inches of heavy wet snow along the path of major eastern cities, newspapers ballyhooed the event as a promise of things to come: "Now may we get out the ulster and overshoes and comforters and furs and ear laps and dance to the merry music of sleigh bells to keep the blood a-tingling. So determined were the elements in this year of grace and Yankee Doodle that we should have a regular old-fashioned winter such as we used to have in the days before open winters and green Christmases."

In the wake of that storm, temperatures descended to unseasonable lows throughout the Midwest and East, as typified by a zero reading in Lexington, Kentucky. After a brief respite on Thanksgiving Day, a new storm from the ocean buried the East in its fleece, beginning on the afternoon of the twenty-sixth. New York City reported a slushy accumulation of 9.7 inches, with seven-foot drifts swirled about by winds over forty miles an hour, breaking all previous records for single-storm depths in the month of November.

It was then that the S.S. *Portland* was overwhelmed by the wind, and waves somewhere in the waters off Provincetown, Massachusetts. No one lived to tell the tale of the circumstance under which it went down, but it is well known that the skipper of the ship knew that a storm of exceptional ferocity would strike him as soon as he reached open water. The public's fascination was stirred with morbid curiosity over the event, and for weeks speculation surfaced over the intentions of the brash young captain. Was he trying to prove the seaworthiness of his ship, or the prowess of his navigational skills? Was he indifferent to the Weather Bureau's warnings? After all, the trustworthiness of their forecasts was often met with mockery in an era prior to satellite photographs and computer printouts. Oftentimes, many a crusty old seaman relied upon "signs" rather than the warnings of a government employee.

Whatever the reason, everyone aboard perished in the most agonizing of conditions, and sea tragedies during this storm were rife, although this was the most spectacular. It was indeed a peculiar time, meteorologically, for at the turn of the month another storm of moderate intensity added to the miseries of eastern America, but the most appalling storm of the season was yet brewing in the western gulf. By December third, a rainstorm of magnificent strength had moved inland to southern Alabama causing drenching rains in the south and middle Atlantic states. The forecasts were ominous: the New York *Times* warned of "Heavy rain or snow—northeast hurricanes." The worst came to pass on the evening of the third as thundering easterly winds, which reached eighty-three miles an hour in New York City, accompanied by pounding rains, turned the considerable remnants of the previous weeks' snow into morasses of slush and raging torrents.

Further to the west, heavy snows blanketed the hills of the Alleghenys from Tennessee north to New England. Most eastern cities were plunged into total darkness and put out of communication with one another, as telegraph wires snapped and dangled as they had so often in the middle and late nineteenth century. The damage was immense—in the Baltimore area alone, more than 800 houses were unroofed, thousands of chimneys were blown down, and trees uprooted. Additionally, a landslide measuring fifteen feet high and more than fifty feet in length mired a train hopelessly in mud in trees, causing it to roll down an embankment. Areas to the west of the storm track picked up between twelve and twenty-eight inches of additional snow, some of the heaviest accumulations known in years in places like Columbus (Ohio), Cincinnati, Pittsburgh, and Buffalo. New York City witnessed its worst fire to date as the Home Life Building in the Wall Street district was completely demolished, as was the neighboring postal building, amidst explosions and suspected arson. Perhaps the only fire of greater devastation in Gotham occurred during an equally noteworthy weather event in December of 1835, during an early season cold wave of unprecedented severity.

Although this account is admittedly a whirlwind tour of how the elements ganged up on the eastern sections of the United States in a relatively short period, it must be underscored that this occurred in a usually tranquil time of year, just after the parting of Indian summer in southern regions and prior to the usual wintry offerings in northern sections. Any one of these cyclones alone would be worthy of front-page notice in any number of eastern chronicles, but their combined effect was nearly unfathomable to those who endured the onslaught of the unseasonable weather. Winter was hardly over at this time, for the great snow and cold of February 1899 was to make history once again, and is described in detail in the chapter discussing snowstorms and blizzards. Other than the period covering 1856-1857, this was truly the standout winter in the latter half of the nineteenth century.

48.
A MEMORABLE THANKSGIVING

ON WEDNESDAY, NOVEMBER 22, 1950, HUNDREDS OF THOUSANDS WERE ON THE MOVE, heading for their holiday destinations. It was a mild day, and little did anyone suspect that a sequence of events would make this Thanksgiving stand out from all the rest. The first thing that went wrong, terribly wrong, was a massive train wreck on the ill-starred Long Island Railroad. Ten cars, packed with commuters and holiday travelers, smashed into an idle train in the Richmond Hill section of Queens. When authorities finally removed everyone from the wreckage, they counted seventy-seven dead and 153 injured. This event alone marred the holiday for thousands of people. As things turned out, it was only the beginning.

The season's first artic blast was already plunging down through central Canada, destined to meet with a warm humid air mass in the East. An immense swirling storm developed, gaining intensity over the western Carolinas and was to wreak havoc for millions that weekend. Temperatures in the Midwest and South began to plunge dramatically. By Saturday morning, Mount Mitchell, the highest point east of the Mississippi, had recorded a low of nineteen degrees below zero, and nearby Ashville, North Carolina, reported two below. Temperatures along the Gold Coast of Florida plunged into the twenties, and Miami notched an even thirty-two degrees. Such was the contrast of these two air masses, adding to the awesome power of the developing storm.

Snow began to fall in swirling masses throughout the southern states of Virginia, North Carolina, Georgia, and Alabama on Friday the twenty-fourth. Around midnight, gale winds enveloped the entire East and Midwest, with drenching rains to the east of the Appalachian Mountains and blizzard conditions to the west. As the storm moved northward into western New York state, it intensified to the point that the Weather Bureau issued a statement that it was "the most violent of its kind ever recorded in the northeastern quarter of the United States." That estimate still stands.

The East Coast at the time had been in the grips of a mini-drought, and airplanes bearing silver-iodine seeds dumped them into the clouds over the Catskills in New York. To say that their operation was successful would be quite an understatement. By 11:30 on the morning of the twenty-fifth, wind swept rains were pounding the East in amounts up to four inches, and the easterly winds had increased in speed to nearly 100 miles an hour, setting records for velocity in New York

City and Boston. The Empire State Building was said to sway 1½ inches from the vertical. By noon, the city of New York "looked as though a giant Halloween prankster had been on a spree," smashing store windows, collapsing buildings, tearing off roofs, and toppling trees. It was, in the words of New Englanders who remembered, "worse than the hurricane of 1938."

Most communities were plunged into darkness, and floods ran as high as five feet in the streets. Falling debris was a hazard, and telephone and power wires dangled overhead like a massive spider web. People who ventured out were literally blown off their feet. High tides and thundering surf smashed against sea walls until they crumbled and flooded coastal highways up and down the East Coast. It was ten days before power was completely restored. One survivor, eighteen-month-old Christianne Citron, was napping when her mother heard the crash of hundreds of pounds of bricks and concrete falling through the roof. She tried to push her way into the child's room, but debris prevented the door from opening. She could hear the child's screams. She enlisted the help of the building superintendent, and the two of them managed to break into the room, where they found the roof, which was ten stories above them, littered all about the room. Miraculously, the baby was safe, and moments after they removed her, another barrage of debris came crashing down, much of it landing in Christianne's crib.

West of Harrisburg, Pennsylvania, it was a different story. Snow shook out of the sky a whole winter's store, and the cities of Cleveland, Pittsburgh, and Erie all set modern snowfall records for a twenty-four-hour period, averaging twenty-seven to thirty inches, much of it swarming into huge drifts by the hurricane-force winds. The snow which began early Friday morning did not stop until some time the following Monday, and thousands of families were at the mercy of flooding streams and rivers. Catastrophe struck in Altoona, Pennsylvania, where the Juniata River poured over its banks, terrorizing a community of 200,000. Meanwhile, thirty-ton Sherman tanks were assigned to push aside the snow, if only to allow the regular snow-removing trucks access to the roads. Clarksburg, West Virginia, picked up forty-two inches of fresh snow, taking top honors in that category.

With sustained winds averaging more than sixty-five miles an hour throughout the region, this appears to be the greatest storm since accurate wind measurements have been taken. At least 273 people lost their lives to this storm, and the accumulated property damage exceeded $500,000,000 over twenty-two states. Falling trees and contact with live wires accounted for most of the casualties in the East, while exposure and exertion took the lives of those in the Snow Belt. November has a history of providing some phenomenal wind storms, and many have disappeared with a shroud of mystery surrounding their fate, but this one apparently topped them all.

49.
FAIR AND CONTINUED COOL

ONE OF THE MOST REMARKABLE EARLY-SEASON SNOWFALLS IN THE EAST WAS ALSO ONE OF its most surprising, especially if one reads the forecast offered in the New York *Times* for that day. "Fair and continued cool" was the official prognostication, confidently uttered from the offices of the U.S. Weather Bureau the evening before. In November of 1953, surveillance methods were adequate for officials to be aware of a slow-moving storm system off the central Florida coastline, but winds aloft had been westerly, and it was the considered opinion of the professionals that this intensifying storm system would be shuttled out to sea no further north than the Carolina coastline.

Within twenty-four hours, history was to prove this judgment flagrantly wrong. One would have been better off consulting the *Old Farmers Almanac*, for they promised a "turbulent and erratic" winter and predicted at least "one spanking northeast storm along the eastern seaboard in the first two weeks of November." Temperatures throughout the mid-autumn season had been moderate, but a large Canadian high-pressure cell was drifting along the northern tier of the country and taking up residence over northern New England, setting a north-to-northeast wind flow and dropping temperatures into the thirties as far south as Washington, D.C. Late in the evening of the fifth of November, the Florida storm headed due north, and the westerly flow aloft dissipated, allowing the coastal disturbance to penetrate the Mid-Atlantic region.

By the morning of the sixth, snow broke out over a wide band, spreading a canopy of white from the Appalachians eastward to a greatly surprised citizenry and to the well-deserved embarrassment of the Weather Bureau. Temperatures dropped steeply from the upper thirties to the middle twenties, when the snow commenced, and new record lows for the date were established from Washington to Boston. The combination of the circulations around the converging high- and low-pressure centers sent the strong northeasterly winds soaring to gale speeds of fifty miles an hour, drifting the early season snowfall into huge blockades. At the coast, a mixture of snow, sleet, and freezing rain left a slushy accumulation of three inches in Atlantic City. Hurricane force winds whipped the waves into monumental thirty-foot towers of destruction. Over 150 square blocks of Atlantic City were submerged under 2½ feet of water, and the Boardwalk was smashed to splinters in this storm that was compared unfavorably to the Hurricane of 1944.

In fact, the entire shoreline from

southern New Jersey to the New England coastal waters was brutally slashed by this monstrous storm, as the wind swept the exceptionally high tides inland like an assaulting legion of the sea itself. Thousands of coastal residents were evacuated, while at Manasquan, New Jersey, a fifty-foot cruiser was swept upriver and hurled against a bridge, bringing the bridge down with it. Many dramatic rescues took place, especially along the Long Island and Connecticut shorelines, with unusually high damage in the area between Stamford and Westport.

Further westward, and to the north, in New England, new early-season snow records were established as twenty-six inches of snow piled up in Upstate New York. Route 17 in New York State was completely shut down, as the absence of snow tires or chains stranded nearly all vehicles. The highway resembled a huge parking lot. Dr. Francis W. Reichelderfer, chief of the United States Weather Bureau, ordered a special study of the storm, to determine how so many forecasters were caught off guard. After the great snowfall of December 26, 1947, in New York City and environs, private weather services began cropping up around the eastern portion of the country, and one of them, the New England Weather Service, correctly gauged the impact of this disturbance and undoubtedly won the confidence of many new clients.

50.
THE WRECK OF THE *ANDREA DORIA*

THE WATERS OFF THE NEW ENGLAND COAST OF NANTUCKET ARE AMONG THE FOGGIEST coastal areas anywhere in the United States, caused by warm land breezes passing over the still cool seas, perpetually chilled by the Labrador current. The area also marks the narrowest passage of ships bound to and inbound from Europe and places beyond. For that reason, a lightship was placed forty miles at sea to guide ships around the dangerous Nantucket shoals, fog or no fog. While no strict rules of passage existed on the night of July 25, 1956, there was an unwritten code among ocean liners that dictated the one-way avenues of the sea, and this is only one of the many mysteries surrounding the wreck of the *Andrea Doria* on that fateful night.

The 29,000-ton *Andrea Doria* from Genoa, Italy, was bound for New York City, and many of the ship's 1,134 passengers were up later than usual the evening of the twenty-fifth, for it was their last night at sea. For many, it was their last night. On a collision course, the Swedish vessel *Stockholm* had departed from New York that morning with a passenger list of 550, and for some reason yet unestablished was ten miles off course, to the north. Both ships were equipped with the most advanced radar equipment of the times, the same technology, in fact, that Weather Service offices were using in the early 1980s. Capable of searching at least ten different ranges through the zero-visibility fog, their radar systems were of little use on that summer evening. Some people claim that, in the dense fog, reflection can render these devices useless in the shorter ranges while others, at the time, disputed this contention. The truth of the matter was unknown that night, the only truth being the awful calamity and subsequent rescue that was at hand.

As the *Andrea Doria* sailed past 40°30' north latitude and 69.50° west longitude, about forty miles to the south of the island of Nantucket, the orchestra was playing "Arrivederci Roma", for maybe the thirtieth time that night, and dancers swayed on the romance that is the seas, perhaps to stay up until the wee hours of the morning. No passenger on that fated ship was to catch a wink of sleep past 11:22 P.M. One reveler reported that she "happened to look out the window and there was this ship. It was right on top of us. A moment later, it struck us. I could see his lights and everything. The rails almost touched ours. It was the *Stockholm*." Forty-five feet of the bow of the *Stockholm* penetrated the starboard side of the *Andrea Doria*, sending her on a twenty-five degree list almost immediately.

The *Andrea Doria* lists on her side, taking on water. (*Department of the Navy*)

Within a minute, both ships sent out an SOS and the captain of the *Stockholm* reported, "bow crushed...badly damaged," while the *Andrea Doria* pleaded, "Danger immediate...need lifeboats, as many as possible...can't use our lifeboats." Water gushed into the starboard side of the wounded *Andrea Doria*, and its startled passengers clung to life on the portside railings. On board were many dignitaries, including Philadelphia mayor Richardson Dilworth. Some described it as a "substantial jar," knocking down tables and drinks and splintering cabins on the starboard deck.

Perhaps the most incredible escape was made by the daughter of the renowned newsman Edward P. Morgan. Fourteen-year-old Linda Morgan had just bedded down for the night when the Swedish ship plowed into the *Andrea Doria*. The bow of the *Stockholm* virtually lifted her out of bed, spatula style, onto its own deck. Somewhat bruised, Linda screamed for help, and when it arrived, they extricated her from the waist-deep scrap that was the bow of the *Stockholm*. Later, Linda was reported missing and presumed dead, and when she was discovered in a Boston hospital the following day, her father reported on the 7 o'clock news that she survived "by what some would call a miracle." Another survivor, Nicola Difiore, had just been through his third shipwreck, seeing two in World War II. Refusing the line's offer of another sea passage to Europe, he intoned, "I never want to see another ship."

The drama played out by marine radio for the next two hours before help arrived. Neighboring ships responded to the crisis, but none was in the immediate vicinity. The first to arrive was the United Fruit Line, which swiftly lowered its eight lifeboats and took on survivors. The *Stockholm*, assessing its seaworthiness, soon took on victims from the *Andrea Doria*. The need for additional lifeboats and assistance,

The *Moonbeam* takes on survivors. (*Department of the Navy*)

however, was urgent. No offer of help, even as little as two lifeboats four hours away, was turned down.

The *Ile de France*, on its way to England with 1,800 aboard, was twenty-five miles north of Nantucket when the SOS was sounded and participated in the rescue for three and one-half hours, until the dawn pink spread across the sky and mirrored the oily waters. Captain Raoul de Beaudeau, the skipper of the *Ile de France*, was credited with saving 753 lives; he heard the first distress signals put out in the fog. "I gave a mental prayer for the clearing of the fog." His prayer was answered as the fog tattered and drifted, and the sight of his crew's faces was ghostly in the glare of the ship's floodlights. A child was born in one of his rescue boats. He saved, in fact, 754 lives. The greatest part of the evacuation effort was completed by 5:00 A.M., and Captain de Beaudeau reported, "All passengers rescued." Well, not quite.

It was not until several days later, when people compared the passenger lists with the survivors, that it was surmised that at least forty-eight people lost their lives, most from the immediate impact of the *Stockholm* smashing into the starboard cabins of the *Andrea Doria*. The *Andrea Doria* was the last word in modern design and comfort, and many had described it as a floating art museum. Indeed, several million dollars' of original artwork went down with her when she sank later that morning. What was comforting from this experience was the immediate and widespread response from all types of passing craft, large and small ships, coast guard and navy vessels, luxury liners and fishing boats. What was not so comforting were the forty-eight dead, who slipped into the ocean virtually unnoticed and probably half asleep and now 250 feet beneath the surface waters with the *Andrea Doria*.

The end came for the Italian liner at 10:09 A.M. on the twenty-sixth, after lying on her side, wounded, for nearly eleven hours. Under the brilliant summer sky, the *Andrea Doria*, with her paint unblemished, her portholes unbroken, and her promenade lights on, gave in to the tremendous pressures of the sea, sending huge geysers into the air when she finally gave up. Only

The Swedish vessel *Stockholm*, though badly damaged, manages to limp back to port. (*Department of the Navy*)

minutes earlier, the captain and crew of the ship slid down her hull when she was considered a total loss. Nearly one thousand former passengers watched her die from the bridge of the *Ile de France* and the *Cape Ann*, a small German vessel that took on 117 former passengers. There was no sound from the rescue ships, only a murmur and the silence of awe. Divers reported that the *Andrea Doria* was lying on the ocean floor on its starboard side, with bubbles of air making the water almost blue; "the *Andrea Doria* is stirring...so completely out of place...that it seems almost alive."

Why the two vessels collided remains a mystery. When the sea is calm, fog is likely in the North Atlantic, and clear weather generates rough seas. Were there radar problems? Was it the fog? Did the high sunspot activity interfere with communications? There were conflicting reports as to whether or not one ship spotted the other, and who was at fault. As usual, the questions were debated in court amidst the many lawsuits that were precipitated by this incident. Nevertheless, this area of the Atlantic has claimed many other ships, under similar circumstances. At the height of the crisis, one of the rescuers cracked, "All we need now is a hurricane." Moments later, over the crackling radio, came word that the season's first hurricane had formed in the Gulf of Mexico only one hour after the collision. Such is life in the dark and fog of the Atlantic's more notorious graveyards.

51. THE GREAT EASTERLY GALE OF 1962

EVERY TWENTY-EIGHT DAYS OR SO, THE PERIOD OF ONE LUNAR CYCLE, THE SUN AND MOON arrive in a position to exert the greatest gravitational force upon the earth and tides. Some forecasters believe that herein lies one of the many clues to respectable long-range forecasting, for the reasoning is that if such a powerful force is sufficient to affect the tides of the world's oceans, why not the atmosphere itself? One of the great storm factories of the world, in the Gulf of Alaska, is notable for spawning a "family" of storms which traverse the Pacific over a period of weeks and subsequently affect the weather of the continental United States. Whether or not there is a correlation between atmospheric "belches" and astronomical influences remains a mystery, but the possibility distinctly exists that there is.

Whatever the cause, a storm of astounding magnitude slowly swept the East Coast in March of 1962 at precisely the worst time—the time of month when the heavenly bodies were exerting their greatest influence upon the tides. From the Carolinas north to New England, a broad unfettered 600-mile fetch of easterly winds developed and lashed the East Coast of the United States for three unforgettable days. Combined with the gale winds attending the storm, this influence piled an extra fathom of water upon the cresting tides and lathered the sea into monumental twenty-to-thirty-foot waves, which cut away at the shoreline from Virginia to Maine.

This mighty storm occurred back in the days when Soviet Premier Khrushchev threatened to "bury us" and Casey Stengel's amazing Mets were opening camp in Ft. Lauderdale, threatening to be buried in the notoriety that becomes a major league team that loses three times as many games as they win in one season. It was in the early days of our involvement in Vietnam; as the *Times* drolly noted, "Washington recognizes the powder keg nature of this situation which could erupt into a major war...but U.S. officials doubt that Communist nations will go that far...."

It was also the days of some spectacular aircraft news. The month of February closed with John Glenn becoming the first astronaut from the United States to orbit the earth, and a few days later, on March first, he was heralded in a veritable blizzard of ticker tape in New York City. Only a few days before, he was circling the earth every ninety minutes, viewing coastlines and continents, and now he was exalted on Wall Street. Wryly he observed, "I don't know who's going to clean up but it was wonderful, wasn't it?" Little did he realize how prophetic his words were. Only an hour before the official cele-

Wave and tidal action undermine houses following the disastrous extratropical storm at Virginia Beach, Virginia. (*National Oceanic and Atmospheric Association*)

bration began, a Boeing 707 plunged into Jamaica Bay, in Queens, exploding into fiery fragments as all 95 aboard perished in the icy waters. It was the highest casualty list on a commercial airline in the history of the country.

The following day, Francis Gary Powers was officially exonerated by the CIA and Congress for his role in the U-2 spy-plane incident over Russia. Only two years earlier, members of Congress had been muttering that he should have "killed himself" rather than being captured and tried by the Russians. The next day, the third of March, a B-58 jet bomber streaked from Los Angeles to New York and back in a record-breaking four hours and forty-two minutes, having made the first leg of the trip in only two hours and one minute. He also left behind a trail of shattered windows, crumbling walls, and startled people. The captain, Robert Sowers, boasted, "If a cannon ball had been fired at the same time we left here (Los Angeles), we would have had time to land in New York and have lunch before it got there."

The very same day, another tragic airplane crash took place, this time over Africa, killing 111, setting yet another new record for civilian casualties. The following day, March fifth, a low-pressure center traveling southeast from a position over Cincinnati merged with a deep-pressure trough off the Mid-Atlantic states to set the east winds blowing. Rain descended heavily in coastal regions, and gobs of wet snow piled up in inland areas, some sections of Virginia receiving upwards of thirty inches. By the early morning hours of the sixth of March, gale winds whistled up and down the coast, reaching a maximum of eighty-four miles an hour in Atlantic City, kicking up thunderous surf, and undermining summer cottages as though they were made of cards.

While the East Coast, as far south as Florida, was involved by this monstrous storm, the shorelines of Delaware, Maryland, and New Jersey were particularly hard hit. Tragedy struck twice on the wave-swept sands of Long Beach Island, just north of Atlantic City.

Severe beach erosion at Rehoboth Beach, Delaware, following the March 1962 storm. (*National Oceanic and Atmospheric Association*)

While evacuation plans were being carried out, a Coast Guard rescue squad drowned while trying to save the lives of dozens of residents, and nine perished. Later in the day, upon the rain-swept streets of Beach Haven, a Coast Guard amphibious vehicle overturned, and those aboard linked hands in the surging white foam and undertow of the sea. Four people were lost to the waves in this incident, forever on the ocean floor of the Atlantic.

In Long Island, David Garroway had just sold his sea-facing home for $39,000 to an unsuspecting New Yorker because the sea spray interfered with his hobby of stargazing. The day after the transaction took place, the waves in Westhampton Beach took the home and forty-nine others out to sea. Up and down the coastline, many areas were made uninhabitable. Chincoteague Island saw all of its residents evacuated, while its most famous resident, a pony named Misty, was safe and sound. Misty had been the subject of concern of children and adults all over the country, for she had been featured in children's books, and later starred in a television series. Long Beach Island was nearly washed off the map, with some ninety percent of its dwellings obliterated or badly damaged by the storm. The governor of Delaware pronounced the event "the worst catastrophe the state of Delaware has ever suffered." The area where the Wright brothers first experimented with flying, Kitty Hawk, North Carolina, on the Outer Banks, suffered severe beach erosion, while 2½ miles of boardwalk was washed to sea in Ocean City, Maryland.

Some ninety miles southeast of Cape Hatteras, a large Liberian oil tanker split up after two days of continuous pounding by the enormous waves. All but one aboard were saved as the captain described the event as "...a big crash, like the rending of metal."

Damage from this storm was set at about $200,000,000 including at least $50,000,000 each in the states of New Jersey, Delaware, and Virginia. It pales in comparison with the great hurricanes of this era, as Diane caused an estimated 1.67 billion dollars' damage and hurricane Agnes exceeded the two-billion-dollar mark. Several days later, the storm spent its fury on the Atlantic coasts of Britain, France, and Ireland, inundating that coastline with forty-foot waves and ninety-five-mile-an-hour winds. So powerful was its influence that it took three days for the winds along the East Coast of America to die down to a frisky breeze, long after the rain and snow had stopped falling.

APPENDIXES

APPENDIX A: EXTREMES OF COLD BY STATE

State	°F	°C	DATE	LOCATION
ALABAMA	-27	-33	1/30/66	New Market
ALASKA	-80	-62	1/23/71	Prospect Creek
ARIZONA	-40	-40	1/7/71	Hawley
ARKANSAS	-29	-34	2/13/05	Pond
CALIFORNIA	-45	-43	1/20/37	Boca
COLORADO	-60	-51	1/1/79	Maybell
CONNECTICUT	-32	-36	1/16/43	Falls Village
DELAWARE	-17	-27	1/17/1893	Millsboro
D.C.	-15	-26	2/11/1899	Washington
FLORIDA	-2	-19	2/13/1899	Tallahassee
GEORGIA	-17	-27	1/27/40	Calhoun
HAWAII	14	10	1/2/61	Halaekale, Maui Is.
IDAHO	-60	-51	1/18/43	Island Park Dam
ILLINOIS	-35	-37	1/22/30	Mount Carroll
INDIANA	-35	-37	2/2/51	Greensburg
IOWA	-47	-44	1/12/12	Washta
KANSAS	-40	-40	2/13/05	Lebanon
KENTUCKY	-34	-37	1/28/63	Cynthiana
LOUISIANA	-16	-27	2/13/1899	Minden
MAINE	-48	-44	1/19/25	Van Buren
MARYLAND	-40	-40	1/13/12	Oakland
MASSACHUSETTS	-34	-37	1/18/63	Birch Hill Dam
MICHIGAN	-51	-46	2/9/34	Vanderbilt
MINNESOTA	-59	-51	1/16/03	Pokegama Dam
MISSISSIPPI	-19	-28	1/30/66	Corinth
MISSOURI	-40	-40	1/13/05	Warsaw
MONTANA	-70	-57	1/20/54	Rogers Pass
NEBRASKA	-47	-44	1/12/1899	Camp Clarke
NEVADA	-50	-46	1/8/37	San Jacinto
NEW HAMPSHIRE	-46	-43	1/28/25	Pittsburg
NEW JERSEY	-34	-37	1/5/04	River Vale
NEW MEXICO	-50	-46	2/1/51	Gavilan
NEW YORK	-52	-47	1/18/79	Old Forge

NORTH CAROLINA	-29	-34	1/30/66	Mt. Mitchell
NORTH DAKOTA	-60	-51	1/15/36	Parshall
OHIO	-39	-39	1/10/1899	Milligan
OKLAHOMA	-27	-33	1/18/30	Watts
OREGON	-54	-48	1/18/30	Seneca
PENNSYLVANIA	-42	-41	1/5/04	Smethport
RHODE ISLAND	-23	-31	1/11/42	Kingston
SOUTH CAROLINA	-20	-29	1/18/77	Longcreek
SOUTH DAKOTA	-58	-50	2/17/36	McIntosh
TENNESSEE	-32	-36	12/30/17	Mountain City
TEXAS	-23	-31	2/8/33	Seminole
UTAH	-50	-46	1/5/13	Strawberry Tunnel
VERMONT	-50	-46	12/30/33	Bloomfield
VIRGINIA	-29	-34	2/10/1899	Monterey
WASHINGTON	-48	-44	12/30/68	Mazama & Winthrop
WEST VIRGINIA	-37	-38	12/30/17	Lewisburg
WISCONSIN	-54	-48	1/24/22	Danbury
WYOMING	-63	-53	2/9/33	Moran

APPENDIX B: EXTREMES OF HEAT BY STATE

STATE	°F	°C	DATE	LOCATION
ALABAMA	112	44	9/5/25	Centerville
ALASKA	100	38	6/27/15	Fort Yukon
ARIZONA	127	53	7/7/05	Parker
ARKANSAS	120	49	8/10/36	Ozark
CALIFORNIA	134	57	7/10/13	Death Valley
COLORADO	118	48	7/11/1888	Bennett
CONNECTICUT	105	41	7/22/26	Waterbury
DELAWARE	110	43	7/21/30	Millsboro
D.C.	106	41	7/20/30	Washington
FLORIDA	109	43	6/29/31	Monticello
GEORGIA	112	44	7/24/52	Louisville
HAWAII	100	38	4/27/31	Pahala
IDAHO	118	48	7/28/34	Orofino
ILLINOIS	117	47	7/14/54	East St. Louis
INDIANA	116	47	7/14/36	Collegeville
IOWA	118	48	7/20/34	Keokuk
KANSAS	121	49	7/24/36	Alton
KENTUCKY	114	46	7/28/30	Greensburg
LOUISIANA	114	46	8/10/36	Plain Dealing
MAINE	105	41	7/10/11	North Bridgton
MARYLAND	109	43	7/10/36	Cumberland & Frederick
MASSACHUSETTS	107	42	8/2/75	Chester & New Bedford
MICHIGAN	112	44	7/13/36	Mio

MINNESOTA	114	46	7/6/36	Moorhead
MISSISSIPPI	115	46	7/29/30	Holly Springs
MISSOURI	118	48	7/14/54	Warsaw & Union
MONTANA	117	47	7/5/37	Medicine Lake
NEBRASKA	118	48	7/24/36	Minden
NEVADA	122	50	6/23/54	Overton
NEW HAMPSHIRE	106	41	7/4/11	Nashua
NEW JERSEY	110	43	7/10/36	Runyon
NEW MEXICO	116	47	7/14/34	Orogrande
NEW YORK	108	42	7/22/26	Troy
NORTH CAROLINA	110	43	8/21/83	Fayetteville
NORTH DAKOTA	121	49	7/6/36	Steele
OHIO	113	45	7/21/34	Gallipolis
OKLAHOMA	120	49	7/26/43	Toshomingo
OREGON	119	48	8/10/1898	Pendleton
PENNSYLVANIA	111	44	7/10/36	Phoenixville
RHODE ISLAND	104	40	8/2/75	Providence
SOUTH CAROLINA	111	44	6/28/54	Camden
SOUTH DAKOTA	120	49	7/5/36	Gann Valley
TENNESSEE	113	45	8/9/30	Perryville
TEXAS	120	49	8/12/36	Seymour
UTAH	116	47	6/29/1892	St. George
VERMONT	105	41	7/4/11	Vernon
VIRGINIA	110	43	7/15/54	Balcony Falls
WASHINGTON	118	48	8/5/61	Ice Harbor Dam
WEST VIRGINIA	112	44	7/10/36	Martinsburg
WISCONSIN	114	46	7/13/36	Wisconsin Dells
WYOMING	114	46	7/12/00	Basin

APPENDIX C: EXTREMES OF PRECIPITATION BY STATE

Snowstorms

State	*Greatest in 24 Hours*	*Greatest Single Storm*
Alabama	19.2 in (49 cm) Florence 12/31-1/1/64	19.5 in (50 cm) Florence 12/31-1/1/64
Alaska	62.0 in (157 cm) Thompson Pass 12/29/55	175.0 in (446 cm) Thompson Pass 12/26-31/55
Arizona	38.0 in (97 cm) Heber R.S. 12/14/67	67.0 in (170 cm) Heber R.S. 12/13-16/67
Arkansas	25.0 in (64 cm) Corning 1/22/18	25.0 in (64 cm) Corning 1/22/18
California	60.0 in (152 cm) Giant Forest 1/18-19/33	189.0 in (480 cm) Shasta Ski Bowl 2/13-19/59
Colorado	75.8 in (193 cm) Silver Lake 4/14-15/21	141.0 in (358 cm) Ruby 3/23-30/1899
Connecticut	28.0 in (71 cm) New Haven 3/12/1888	50.0 in (127 cm) Middletown 3/11-14/1888
Delaware	25.0 in (64 cm) Dover 2/19/79	25.0 in (64 cm) Dover 2/19/79
Florida	4.0 in (10 cm) Milton Exp. Station 3/6/54	4.0 in (10 cm) Milton 3/6/54
Georgia	19.3 in (49 cm) Cedartown 3/2-3/42	19.3 in (49 cm) Cedartown 3/2-3/42
Idaho	38.0 in (97 cm) Sun Valley 2/11/59	60.0 in (152 cm) Roland W. Portal 12/25-27/37
Illinois	36.0 in (91 cm) Astoria 2/27-28/1900	37.8 in (96 cm) Astoria 2/27-28/1900
Indiana	20.0 in (51 cm) La Porte 2/12/44*	37.0 in (94 cm) La Porte 2/14-19/58
Iowa	21.0 in (53 cm) Sibley 2/18/62	30.8 in (78 cm) Rock Rapids 2/17-21/62
Kansas	26.0 in (66 cm) Fort Scott 12/28-29/54	37.0 in (94 cm) Olathe 3/23-24/12
Kentucky	18.0 in (46 cm) Bowling Green 3/9/60 and Celicia 11/2/66	27.0 in (69 cm) Bowling Green 3/7-11/60
Louisiana	24.0 in (61 cm) Rayne 2/14-15/1895	24.0 in (61 cm) Rayne 2/14-15/1895
Maine	35.0 in (89 cm) Middle Dam 11/23/43	56.0 in (142 cm) Long Falls Dam 2/24-28/69
Maryland	31.0 in (79 cm) Clear Spring 3/29/42	36.0 in (91 cm) Edgemont 3/29-30/42
Massachusetts	28.2 in (72 cm) Blue Hill, Milton 2/24-25/69	47.0 in (119 cm) Peru 3/2-5/47

Greatest Month	*Greatest Season*
24.0 in (61 cm) Valley Head 1/40	25.0 in (64 cm) Valley Head 1939-40
297.9 in (757 cm) Thompson Pass 2/53	974.5 in (2475 cm) Thompson Pass 1952-53
104.8 in (266 cm) Flagstaff 1/49	226.7 in (576 cm) Hawley Lake 1967-68
48.0 in (122 cm) Calico Rock 1/18	61.0 in (155 cm) Hardy 1917-18
390.0 in (991 cm) Tamarack 1/11	884.0 in (2245 cm) Tamarack 1906-07
249.0 in (632 cm) Ruby 3/1899	779.0 in (1979 cm) Ruby 1896-97
73.6 in (187 cm) Norfolk 3/56	177.4 in (451 cm) Norfolk 1955-56
36.0 in (91 cm) Milford 2/1899	49.5 in (126 cm) Wilmington 1957-58
4.0 in (10 cm) Milton 3/54	4.0 in (10 cm) Milton 1953-54
26.5 in (67 cm) Diamond 2/1895	39.0 in (99 cm) Diamond 1894-95
143.8 in (365 cm) Burke 1/54	441.8 in (1122 cm) Roland W. Portal 1949-50
47.0 in (119 cm) Astoria 2/1900	77.0 in (196 cm) Chicago 1969-70
59.8 in (152 cm) La Porte 2/58	122.3 in (311 cm) La Porte 1962-63
42.0 in (107 cm) Osage, Northwood 3/15	90.4 in (230 cm) Northwood 1908-09
55.9 in (142 cm) Olathe 3/12	100.1 in (254 cm) Goodland 1979-80
46.5 in (118 cm) Benham 3/60	108.2 in (275 cm) Benham 1959-60
24.0 in (61 cm) Rayne 2/1895	24.0 in (61 cm) Rayne 1894-95
88.3 in (224 cm) Long Falls Dam 2/69	238.5 in (606 cm) Long Falls Dam 1968-69
58.0 in (147 cm) Oakland 1/1895	174.9 in (444 cm) Deer Park 1901-02
78.0 in (198 cm) Monroe 2/1893	162.0 in (411 cm) Monroe 1892-93

Michigan	27.0 in (69 cm) Dunbar 3/29/47 and Ishpemig 10/23/29	46.1 in (117 cm) Calumet 1/15-20/50
Minnesota	28.0 in (71 cm) Pigeon R. Bridge 4/4-5/33	35.2 in (89 cm) Duluth 12/5-8/50
Mississippi	18.0 in (46 cm) Mt. Pleasant 12/23/63 and Tunica 12/23/63	18.0 in (46 cm) Mt. Pleasant 12/23/63
Missouri	27.6 in (70 cm) Neosho 3/16-17/70	27.6 in (70 cm) Neosho 3/16-17/70
Montana	44.0 in (112 cm) Summit 1/20/72	57.0 in (145 cm) Summit 1/19-21/72
Nebraska	24.0 in (61 cm) Hickman 2/11/65	41.0 in (104 cm) Chadron 1/2-4/49
Nevada	25.0 in (64 cm) Mt. Rose Resort 1/20/69	75.0 in (191 cm) Mt. Rose Resort 1/18-22/69
New Hampshire	56.0 in (142 cm) Randolph 11/22-23/43	77.0 in (196 cm) Pinkham Notch 2/24-28/69
New Jersey	29.7 in (75 cm) Long Branch 12/26-27/47	34.0 in (87 cm) Cape May 2/11-14/1899
New Mexico	30.0 in (76 cm) Sandia Crest 12/29/58	40.0 in (102 cm) Corona 12/14-16/59
New York	54.0 in (137 cm) Barnes Corner 1/9/76	69.0 in (175 cm) Watertown 1/18-22/40
North Carolina	31.0 in (79 cm) Nashville 3/2/27	31.0 in (79 cm) Nashville 3/2/27
North Dakota	24.0 in (61 cm) Lisbon 2/15/15 and Berthold Agency 2/25/30	35.0 in (89 cm) Lisbon 2/13-15/15
Ohio	20.7 in (53 cm) Youngstown 11/24-25/50	36.3 in (91 cm) Steubenville 11/24-26/50
Oklahoma	23.0 in (58 cm) Buffalo 2/21/71	36.0 in (91 cm) Buffalo 2/21-22/71
Oregon	37.0 in (94 cm) Crater Lake 1/17/51**	95.0 in (241 cm) Crater Lake 1/15-19/51
Pennsylvania	38.0 in (97 cm) Morgantown 3/20/58	50.0 in (127 cm) Morgantown 3/19-21/58
Rhode Island	34.0 in (86 cm) Foster 2/8-9/45	34.0 in (86 cm) Foster 2/8-9/45
South Carolina	24.0 in (61 cm) Rimini 2/9-10/73	28.9 in (73 cm) Caesar's Head 2/15-16/69
South Dakota	38.0 in (97 cm) Dumont 3/27/50	60.0 in (152 cm) Dumont 3/26-28/50
Tennessee	22.0 in (56 cm) Morristown 3/9/60	28.0 in (71 cm) Westbourne 2/19-21/60

115.3 in (293 cm) Calumet 1/50	298.3 in (758 cm) Herman 1868-69
66.4 in (169 cm) Collegeville 3/65	147.5 in (375 cm) Pigeon R. Bridge 1936-37
23.0 in (58 cm) Cleveland 1/66	25.2 in (64 cm) Senatobia 1967-68
47.5 in (121 cm) Poplar Bluff 1/18	70.3 in (179 cm) Maryville 1911-12
131.1 in (333 cm) Summit 1/72	406.5 in (1032 cm) Kings Hill 1958-59
59.6 in (157 cm) Chadron 1/49	104.9 in (266 cm) Kimball 1958-59
124.0 in (315 cm) Mt. Rose Resort 1/69	323.0 in (820 cm) Mt. Rose Resort 1968-69
130.0 in (330 cm) Pinkham Notch 2/69*	323.0 in (820 cm) Pinkham Notch 1968-69
50.1 in (127 cm) Freehold 12/1880	108.1 in (275 cm) Culvers Lake 1915-16
144.0 in (366 cm) Anchor Mine 3/12	483.0 in (1227 cm) Anchor Mine 1911-12
192.0 in (488 cm) Bennett Bridge 1/78	466.9 in (1186 cm) Hooker 1976-77
56.5 in (144 cm) Boone 3/60	100.7 in (256 cm) Banner Elk 1959-60
45.5 in (116 cm) Tagus 4/70	99.9 in (254 cm) Pembina 1906-07
69.5 in (177 cm) Chardon 12/62	161.5 in (410 cm) Chardon 1959-60
39.5 in (100 cm) Buffalo 2/71	87.3 in (222 cm) Beaver 1911-12
256.0 in (650 cm) Crater Lake 1/33	879.0 in (2233 cm) Crater Lake 1932-33
86.0 in (218 cm) Blue Knob 12/1890	225.0 in (572 cm) Blue Knob 1890-91
62.0 in (157 cm) Foster 3/56	122.6 in (311 cm) Foster 1947-48
33.9 in (86 cm) Caesar's Head 2/69	60.3 in (153 cm) Caesar's Head 1968-69
94.0 in (239 cm) Dumont 3/50	258.2 in (656 cm) Lead 1976-77
39.0 in (99 cm) Mountain City 3/60	75.5 in (192 cm) Mountain City 1959-60

*Figures have been exceeded at mountaintop stations.

**Also on earlier dates at the same or other places in the state.

Rains

State	Greatest in 24 Hours	Greatest Monthly
Alabama	20.33 in (516 mm) Axis 4/13/55	34.86 in (885 mm) Robertsdale 7/16
Alaska	14.84 in (377 mm) Little Port Walter 12/6/64	70.99 in (1803 mm) Mac Leod Harbor 11/76
Arizona	11.40 in (291 mm) Workman Creek 1 9/4-5/70	16.95 in (431 mm) Crown King 8/51
Arkansas	12.00 in (305 mm) Arkadelphia 6/28/05	23.86 in (606 mm) El Dorado 12/31
California	26.12 in (663 mm) Hoegees Camp 1/22-23/43	71.54 in (1815 mm) Helen Mine 1/09
Colorado	11.08 in (256 mm) Holly 6/17/65	23.28 in (591 mm) Ruby 2/1897
Connecticut	12.12 in (308 mm) Hartford 8/18-19/55	27.70 in (704 mm) Torrington 2 8/55
Delaware	7.83 in (199 mm) Odessa 6/27/38	17.69 in (449 mm) Bridgeville 8/67
District of Columbia	7.31 in (186 mm) 24th & M Streets 8/11/29	17.45 in (443 mm) 24th & M Streets 9/34
Florida	38.70 in (983 mm) Yankeetown 9/5/50	42.33 in (1075 mm) Ft. Lauderdale 10/65
Georgia	18.00 in (457 mm) St. George 8/28/11	30.23 in (768 mm) Blakely 7/16
Hawaii	38.00 in (965 mm) Kilauea Plantation 1/24-25/56	107.00 in (2718 mm) Kukui, Maui 3/42
Idaho	7.17 in (182 mm) Rattlesnake Creek 11/23/09	28.23 in (717 mm) Roland 12/23
Illinois	16.54 in (420 mm) East St. Louis 6/14/57	20.03 in (509 mm) Monmouth 9/11
Indiana	10.50 in (267 mm) Princeton 8/6/05	21.39 in (543 mm) Evans Landing 1/37
Iowa	16.70 in (424 mm) Decatur Co. 8/5-6/59	22.18 in (563 mm) Red Oak 6/67
Kansas	12.59 in (320 mm) Burlington 5/31/41	24.56 in (624 mm) Ft. Scott 6/1845
Kentucky	10.40 in (264 mm) Dunmor 6/28/60	22.97 in (583 mm) Earlington 1/37
Louisiana	22.00 in (559 mm) Hackberry 8/29/62	37.99 in (965 mm) Lafayette 8/40
Maine	8.05 in (204 mm) Brunswick 9/11/54	17.75 in (451 mm) Brunswick 11/1845
Maryland	14.75 in (375 mm) Jewell 7/26/1897	20.35 in (517 mm) Leonardtown 7/45

Greatest Annual	*Least Annual*
106.57 in (2707 mm)	22.00 in (559 mm)
Mt. Vernon Barrack 1853	Primrose Farm 1954
332.29 in (8440 mm)	1.61 in (41 mm)
Mac Leod Harbor 1976	Barrow 1935
58.92 in (1497 mm)	0.07 in (1.8 mm)
Hawley Lake 1978	Davis Dam 1956
98.55 in (2503 mm)	19.11 in (485 mm)
Newhope 1957	Index 1936
153.54 in (3900 mm)	0.00 in (0 mm)
Monumental 1909	Death Valley 1929
92.84 in (2358 mm)	1.69 in (43 mm)
Ruby 1897	Buena Vista 1939
78.53 in (1995 mm)	23.60 in (599 mm)
Burlington Dam 1955	Baltic 1965
72.75 in (1848 mm)	21.38 in (543 mm)
Lewes 1948	Dover 1965
61.33 in (1558 mm)	18.79 in (477 mm)
24th & M Streets 1889	Washington 1826
112.43 in (2856 mm)	22.45 in (573 mm)
Wewahitchka 1966	Key West 1961
122.16 in (3103 mm)	17.14 in (435 mm)
Flat Top 1959	Swainsboro 1954
578.00 in (14,681 mm)	0.19 in (5 mm)
Kukui, Maui, 1931	Kawaihae 1953
81.05 in (2059 mm)	2.09 in (53 mm)
Roland 1933	Grand View 1947
74.58 in (1894 mm)	16.59 in (421 mm)
New Burnside 1950	Keithsburg 1956
97.38 in (2473 mm)	18.67 in (474 mm)
Marengo 1890	Brookville 1934
74.50 in (1892 mm)	12.11 in (308 mm)
Muscatine 1851	Cherokee 1958
65.87 in (1673 mm)	4.77 in (121 mm)
Mound City 1951	Johnson 1956
79.68 in (2024 mm)	14.51 in (368 mm)
Russellville 1950	Jeremiah 1968
111.28 in (2827 mm)	26.44 in (672 mm)
Morgan City 1946	Shreveport 1936
75.64 in (1921 mm)	23.06 in (586 mm)
Brunswick 1845	Machias 1930
72.59 in (1844 mm)	17.76 in (451 mm)
Salisbury 1948	Picardy 1930

Massachusetts	18.15 in (461 mm) Westfield 8/18-19/55	26.85 in (682 mm) Westfield 8/55
Michigan	9.78 in (248 mm) Bloomingdale 9/1/14	16.24 in (413 mm) Battle Creek 6/1883
Minnesota	10.75 in (273 mm) Mahnomen 7/20/09	16.52 in (420 mm) Alexandria 8/1900
Mississippi	15.68 in (398 mm) Columbus 7/9/68	30.75 in (781 mm) Merrill 7/16
Missouri	18.18 in (462 mm) Edgerton 7/20/65	25.54 in (649 mm) Joplin 5/43
Montana	11.50 in (292 mm) Circle 6/20/21	16.79 in (426 mm) Circle 6/21
Nebraska	13.15 in (334 mm) York 7/8/50	20.00 in (508 mm) Tecumseh 6/1883
Nevada	7.40 in (188 mm) Lewers Ranch 3/18/07	33.03 in (839 mm) Mt. Rose 12/64
New Hampshire	10.38 in (264 mm) Mt. Washington 2/10-11/70	25.56 in (649 mm) Mt. Washington 2/69
New Jersey	14.81 in (376 mm) Tuckerton 8/19/39	25.98 in (660 mm) Paterson 9/1881
New Mexico	11.28 in (287 mm) Lake Maloya 5/19/55	16.21 in (412 mm) Portales 5/41
New York	11.17 in (284 mm) NYC Central Park 10/9/03	25.27 in (642 mm) West Shokan 10/55
North Carolina	22.22 in (564 mm) Altapass 7/15/16	37.40 in (950 mm) Gorge 7/16
North Dakota	7.70 in (196 mm) McKinney 6/15/1897	14.01 in (356 mm) Mohall 6/44
Ohio	10.51 in (267 mm) Sandusky 7/12/66	16.13 in (410 mm) Deamos 7/1896
Oklahoma	15.50 in (394 mm) Sapulpa 9/3-4/40	23.95 in (608 mm) Miami 5/43
Oregon	10.17 in (258 mm) Glenora 12/21/15	50.20 in (1271 mm) Glenora 11/09
Pennsylvania	34.50 in (876 mm)* Smethport 7/17/42	23.66 in (601 mm) Mt. Pocono 8/55
Puerto Rico	23.00 in (584 mm) Adjuntas 8/8/1899	52.02 in (1321 mm) Jayoya 10/70
Rhode Island	12.13 in (308 mm) Westerly 9/16-17/32	15.00 in (381 mm) Rocky Hill 8/55
South Carolina	13.25 in (337 mm) Effingham 7/15/16	31.13 in (791 mm) Kingtree 7/16
South Dakota	8.00 in (203 mm) Elk Point 9/10/1900	18.61 in (473 mm) Deadwood 5/46
Tennessee	11.00 in (279 mm) McMinnville 3/28/02	23.90 in (601 mm) McKenzie 1/37
Texas	38.20 in (970 mm)* Thrall 9/9-10/21	34.85 in (885 mm) McKinney 5/1881

70.33 in (1786 mm)	Westfield 1955	21.76 in (553 mm)	Chatham Life Station 1965
64.01 in (1626 mm)	Adrian 1881	15.64 in (397 mm)	Croswell 1936
51.53 in (1309 mm)	Grand Meadow 1911	7.81 in (198 mm)	Angus 1936
102.89 in (2613 mm)	Beaumont 1961	25.97 in (660 mm)	Yazoo City 1936
92.77 in (2356 mm)	Portageville 1957	16.14 in (410 mm)	La Belle 1956
55.51 in (1410 mm)	Summit 1953	2.97 in (75 mm)	Belfry 1960
64.52 in (1639 mm)	Omaha 1869	6.30 in (160 mm)	Hull 1931
50.03 in (1499 mm)	Mt. Rose 1969	Trace	Hot Springs 1898
130.14 in (3306 mm)	Mt. Washington 1969	22.31 in (567 mm)	Bethlehem 1930
85.99 in (2184 mm)	Paterson 1882	19.85 in (504 mm)	Canton 1965
62.45 in (1586 mm)	White Tail 1941	1.00 in (25 mm)	Hermanas 1910
82.06 in (2084 mm)	Wappingers Falls 1903	17.64 in (448 mm)	Lewiston 1941
129.60 in (3291 mm)	Rosman 1964	22.69 in (576 mm)	Mount Airy 1930
37.98 in (965 mm)	Milnor 1944	4.02 in (102 mm)	Parshall (near) 1934
70.82 in (1799 mm)	Little Mountain 1870	16.96 in (431 mm)	Elyria 1963
84.47 in (2146 mm)	Kiamichi Tower 1957	6.53 in (166 mm)	Regnier 1956
168.88 in (4290 mm)	Valsetz 1937	3.33 in (85 mm)	Warmspring Reservoir 1939
81.64 in (2074 mm)	Mt. Pocono 1952	15.71 in (399 mm)	Breezewood 1965
253.79 in (6446 mm)	La Mina El Yunguc 1936	9.76 in (248 mm)	Rio Jueyes 1967
65.91 in (1674 mm)	Pawtucket 1888	24.08 in (612 mm)	Block Island 1965
101.65 in (2582 mm)	Caesar's Head 1961	20.73 in (527 mm)	Rock Hill 1954
48.42 in (1230 mm)	Deadwood 1946	2.89 in (48 mm)	Ludlow 1936
114.88 in (2918 mm)	Haw Knob 1957	25.23 in (641 mm)	Halls 1941
109.38 in (2778 mm)	Clarkville 1873	1.64 in (42 mm)	Presidio 1956

State	*Greatest in 24 Hours*	*Greatest Single Storm*
Utah	6.00 in (152 mm) Bug Point 9/5/70	19.14 in (486 mm) Buckboard Flat 10/72
Vermont	8.77 in (223 mm) Somerset 11/3-4/27	16.99 in (432 mm) Mays Mill 10/55
Virginia	27.00 in (686 mm)* Nelson County 8/20/69	23.88 in (607 mm) Big Meadows 8/55
Washington	12.00 in (305 mm) Quinault Ranger Station 1/21/35	57.04 in (1449 mm) Peterson's Ranch 12/33
West Virginia	19.00 in (483 mm)* Rockport 7/18/1889	16.30 in (414 mm) Princeton 6/01
Wisconsin	11.72 in (298 mm) Mellen 6/24/46	17.41 in (442 mm) Hayward 8/41
Wyoming	5.50 in (140 mm) Dull Center 5/31/27	12.78 in (325 mm) Alva 5/62

Greatest Month	*Greatest Season*
70.71 in (1796 mm)	1.34 in (34 mm)
Alta 1975	Myton 1974
73.61 in (1870 mm)	22.98 in (584 mm)
Mt. Mansfield 1969	Burlington 1941
81.78 in (2077 mm)	12.52 in (318 mm)
Montebello 1972	Moores Creek Dam 1941
184.56 in (4688 mm)	2.61 in (66 mm)
Wynoochee Oxbow 1931	Wahluke 1930
94.01 in (2388 mm)	9.50 in (241 mm)
Romney 1948	Upper Tract 1930
62.07 in (1577 mm)	12.00 in (305 mm)
Embarrass 1884	Plum Island 1937
55.46 in (1409 mm)	1.28 in (33 mm)
Grassy Lake Dam 1945	Lysite 1960

***Estimated from hydrologic bucket survey.**

INDEX BY PRINCIPAL LOCALE

United States Locations

Canadian Locations